Electric Motors and Control Systems, 2e

Electric Motors and Control Systems, 2e

Activities Manual

Frank D. Petruzella

ACTIVITIES MANUAL FOR ELECTRIC MOTORS AND CONTROL SYSTEMS,
SECOND EDITION

7 8 9 10 11 12 QVS/QVS 22 21 20 19 18

ISBN 978-1-259-33283-8
MHID 1-259-33283-7

Senior Vice President, Products & Markets: *Kurt L. Strand*
Vice President, General Manager, Products & Markets: *Marty Lange*
Vice President, Content Design & Delivery: *Kimberly Meriwether David*
Director: *Thomas Timp*
Global Brand Manager: *Raghu Srinivasan*
Director, Product Development: *Rose Koos*
Product Developer: *Vincent Bradshaw*
Marketing Manager: *Nick McFadden*
Director, Content Design & Delivery: *Linda Avenarius*
Executive Program Manager: *Faye M. Herrig*
Content Project Managers: *Jessica Portz, Tammy Juran, and Sandra Schnee*
Buyer: *Jennifer Pickel*
Content Licensing Specialists: *Lenny Behnke*
Cover Image: © *Deyan Georgiev Authentic collection/Alamy RF (background); Photos
courtesy Rockwell Automation, www.rockwellautomation.com (foreground)*
Compositor: *MPS Limited*
Typeface: *11/19 Times New Roman*
Printer: *Quad/Graphics*

www.mhhe.com

Contents

Preface

This manual contains quizzes, practical assignments, and computer-generated simulated circuit analysis assignments.

Quizzes made up of multiple choice, True/False, and completion type questions are provided for each part of each chapter. These serve as an excellent review of the material presented.

Practical assignments are designed to give the student an opportunity to apply the information covered in the text in a hands-on motor installation.

The Constructor motor control simulation software can be downloaded from CMH Software Inc., using the link and access code available with this text. (Note: Those using an electronic copy of this text can access the software at http://highered.mheducation.com:80/sites/0073373818.) This state-of-art program contains some 45 pre-constructed simulated motor control circuits in both NEMA and IEC format. *The Constructor* analysis assignments provide students with the opportunity to test and troubleshoot the motor control circuits discussed in the text. *The Constructor* simulation engine visually displays power flow to each component and using animation and sound effects, each component will react accordingly once power is supplied.

I would like to thank the team from CMH Software Inc.—Bob Hosea, Kevin Christensen, and Jason Lucas—for their collaborative effort in the development of the simulated motor control circuit assignments. I would also like to thank Product Developer Vincent Bradshaw on this, and Don Pelster from Nashville State Community College for the excellent job he did editing the Activities Manual.

Frank D. Petruzella

Electric Motors and Control Systems, 2e

1 Safety in the Workplace

PART 1   Protecting against Electric Shock

Place the answers in the space provided.

1. The higher the body resistance, the greater the potential electric shock hazard.
   (True/False) _____

2. Moisture _____ (increases/decreases) resistance.

3. Generally any voltage above _____ V is considered dangerous.

4. An electric shock with a current of 100 mA is rarely fatal.
   (True/False) _____

5. The pathway taken by electricity through the body has little influence on the
   severity of an electric shock. (True/False) _____

6. An electric shock can result any time a body becomes part of an electric

   _____.

7. The most common electricity-related injury is a burn.
   (True/False) _____

8. The three electrical factors involved in an electric shock are _____,
   _____, and _____.

9. Which of the following statements about electric shock is *not* true?

   a. The lower the body resistance, the greater the potential shock hazard.

   b. The higher the voltage, the greater the potential shock hazard.

   c. The lower the current, the greater the potential shock hazard.

   d. The longer the length of time of exposure, the greater the potential shock hazard.

   Answer _____

10. The amount of current flowing through a body is equal to

    a. *E/R*

    b. *E × R*

    c. *P/R*

    d. *P × R*

    Answer _____

11. _____ percent of electrical workplace accidents are caused by arc flash.

    a. 80

    b. 50

    c. 25

    d. 5

    Answer _____

12. Which of the following is not a common cause of arc flash?

    a. Build-up of conductive dust and dirt

    b. Inadequate personal protection equipment

    c. A dropped tool

    d. Accidental contact with electrical systems

    Answer _____

13. Activities with high potential for exposure to arc flash include

    a. Troubleshooting live circuits

    b. Opening or closing circuits

    c. Grounding circuits

    d. All of these

    Answer _____

14. The primary method for preventing employees from arc flash is to deenergize live parts prior to working on them using proper lockout/tagout procedures. (True/False) _____

15. You should avoid wearing _____ (polyester/cotton) clothing on the job.

16. Leather gloves are worn over rubber electrical gloves to protect the rubber glove from punctures. (True/False)_____

17. Hard hats are made of a nonconductive material as an added protection against electric shock. (True/False) _____

18. Receiving an electric shock is a clear warning that proper safety measures have not been followed. (True/False) _____

19. Keep material or equipment at least _____ feet away from high-voltage overhead power lines.

20. When installing new machinery, ensure that all metal framework is efficiently and permanently _____.

21. Capacitors should be _____ before handling them.

PART 2   Grounding—Lockout—Codes

Place the answers in the space provided.

1. _____ is the intentional connection of a current-carrying conductor to the earth.

2. _____ is the permanent joining together of all non-current-carrying conductive enclosures.

3. The purpose of grounding is to

    a. Limit voltage surges

    b. Provide a ground reference that stabilizes

    c. Facilitate the operation of overcurrent devices

    d. All of these

    Answer _____

4. Bonding establishes a low-resistance path for ground fault current. (True/False) _____

5. Bonding limits the touch voltage when non-current-carrying metal parts are inadvertently energized. (True/False) _____

6. The resistance of the earth as an electrical conductor is very low. (True/False) _____

7. The two distinct parts of a grounding system are the _____ grounding and the _____ grounding.

8. System grounding is the connection of one of the current-carrying conductors to ground. (True/False) _____

9. Which conductors form part of the grounding system?

    a. Equipment grounding conductor

    b. Grounded conductor

    c. Grounding electrode conductor

    d. All of these

    Answer _____

10. A ground fault is an unintentional connection between an _____ conductor of an electric circuit and metallic enclosures and raceways.

11. A ground fault circuit interrupter (GFCI) is a device that senses small currents to _____ and acts to quickly shut off the current.

12. A GFCI is capable of reacting to leakage current as small as _____ mA.

13. A GFCI protects against
    a. Line-to-line short circuit
    b. Ground fault
    c. Overload
    d. All of these
    Answer _____

14. A GFCI monitors
    a. The current flow in the live conductor
    b. The current flow in the neutral conductor
    c. Any difference in current flow between the live and neutral conductors
    d. The sum of the current flow in the live and neutral conductor
    Answer _____

15. Electrical lockout refers to the process of _____ the power source in the off position.

16. Electrical _____ is the process of placing a danger tag on the source of electrical power, which indicates that the equipment may not be operated until the danger tag is removed.

17. Electrical lockout/tagout is at times used in servicing electrical equipment that requires the power to be on. (True/False) _____

18. "Dead" circuits should be
    a. Checked by momentarily touching them
    b. Checked by metering the voltage
    c. Locked out and tagged out before starting any repairs
    d. Both b and c
    Answer _____

19. A lockout/tagout must not be removed by any person other than the authorized person who installed it. (True/False) _____

20. A disconnect switch should not be operated if the switch is still under load. (True/False) _____

21. There should be only one lock and tag on the disconnect switch even if more than one person is working on the machinery. (True/False) _____

22. OSHA, the Occupational Safety and Health Administration, is a federal agency whose mission is to ensure the safety and health concerns of all American workers. (True/False) _____

23. The National Electrical Code (NEC)

    a. Outlines the minimum requirements that must be met for a safe wiring installation

    b. Outlines the ideal requirements for a wiring installation

    c. Is a manual designed to teach how to properly wire houses

    d. Both a and c

    Answer _____

24. Article _____ of the NEC deals with motors and all associated branch circuits.

25. _____ locations include petroleum refineries and gasoline storage and dispensing areas.

26. Use a class _____ –rated fire extinguisher on energized electrical wiring or equipment.

    a. A

    b. B

    c. C

    d. D

    Answer _____

27. In case of fire in the workplace, it is your responsibility to

    a. Know the safest means of exit from the building

    b. Leave the danger area in an orderly manner

    c. Trigger the nearest fire alarm

    d. All of these

    Answer _____

28. _____ or _____ indicates that a piece of electrical equipment or material has been tested and evaluated for the purpose for which it is intended to be used.

29. Which of the following is a recognized testing laboratory?

    a. UL

    b. NFPA

    c. NEC

    d. All of these

    Answer _____

30. Which of the following develops standard frame sizes for motors?

    a. NEMA

    b. IEC

    c. IEEE

    d. Both a and b

    Answer _____

2 Understanding Electrical Drawings

PART 1   Symbols—Abbreviations—Ladder Diagrams

Place the answers in the space provided.

1. In a ladder diagram, components are arranged according to their physical position in the installation. (True/False) _____

2. When conductors cross but make no electrical contact, this is represented by intersecting lines with no dot. (True/False) _____

3. Circuits are normally represented on diagrams in their deenergized position. (True/False) _____

4. The conductors feeding current to a large motor would be considered to be part of the power circuit. (True/False) _____

5. When more than one electrical load is connected in a rung of a ladder diagram, the loads are normally connected in series. (True/False) _____

6. All additional stop or off control devices must be wired in parallel. (True/False) _____

7. Identify each of the symbols shown in Figure 2-1:

a. _____   b. _____   c. _____

7

d. _____   e. _____   f. _____

g. _____   h. _____   i. _____

Figure 2-1   Electrical symbols for Question 7.

8. State a commonly used abbreviation for each of the following terms found on electrical diagrams:

 a. Alternating current _____

 b. Control relay _____

 c. Ground _____

 d. Three-phase motor terminal connections _____

 e. Forward _____

 f. Solenoid _____

 g. Pilot light _____

 h. Motor _____

 i. Normally open _____

 j. Overload relay _____

 k. Breaker _____

l. Rectifier _____

m. Positive _____

n. Push button _____

o. Rheostat _____

p. Secondary _____

q. Transformer _____

r. Time delay _____

s. Phase _____

t. Manual _____

9. On the ladder diagram shown in Figure 2-2, complete the identification of the rungs and the wire numbers.

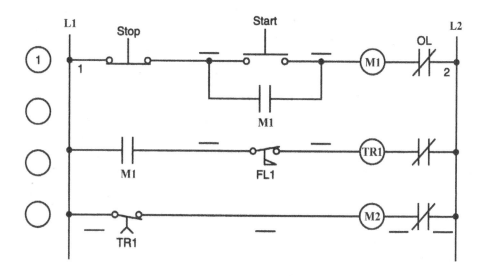

Figure 2-2   Ladder diagram for Question 9.

10. The _____ is the electrical device in a ladder diagram that uses the power supplied from L1 to L2.

11. At least one load must be included in each rung of a ladder diagram. (True/False) _____

12. A broken connection on a ladder diagram indicates a _____ function.

PART 2   Wiring—Single-Line Diagrams—Block Diagrams

1. A wiring diagram is intended to show the actual _____ of all components.

2. Wiring diagrams are used in installing and tracing wires in electrical installations. (True/False) _____

3. Frequently, there will be a wiring diagram inside a magnetic motor starter enclosure. (True/False) _____

4. Each block of a block diagram represents electrical circuits that perform a specific _____.

5. The single-line diagram uses blocks along with a single line to show all the major components of a system. (True/False) _____

6. A conduit layout diagram normally includes the number and size of wires to be run in the conduit. (True/False) _____

7. Which of the following diagrams shows an electrical installation reduced to its simplest form?
   a. Wiring diagram
   b. Ladder diagram
   c. Single-line diagram
   d. Pictorial diagram
   Answer _____

8. On a block diagram, the arrows connecting the blocks indicate
   Answer _____

9. Answer each of the following with reference to the motor diagram shown in Figure 2-3 (p. 11):

Figure 2-3   Motor diagram for Question 9.

a. The two types of motor diagrams shown are a _____ diagram and a _____ diagram.

b. Assuming the incoming line voltage is three-phase, 230 V, both the M coil and the pilot light would be rated for _____ V.

c. The auxiliary M contact will be closed whenever the motor is running. (True/False) _____

d. Assume the overload contact has tripped open. Pressing the start button closed will turn the pilot light on but the starter coil will not energize. (True/False) _____

PART 3   Motor Terminal Connections

1. Motors are classified according to the type of power used as being

   _____ or _____.

2. For the DC motor connections shown in Figure 2-4, identify the motor type and motor leads.

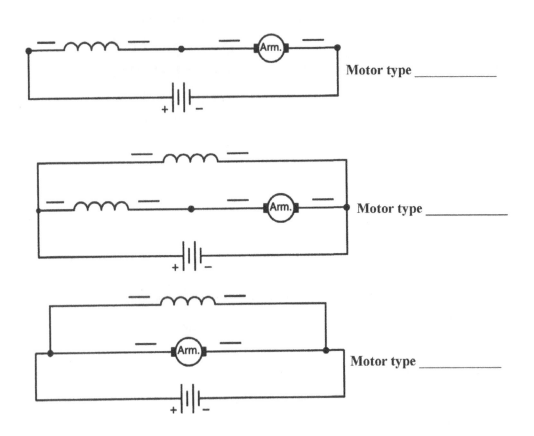

Figure 2-4   DC motor diagrams for Question 2.

3. Industrial applications use DC motors because the

   _____/_____ relationship can be easily varied.

4. Dynamic braking can be used with DC motors to bring the motor to a stop without the aid of a mechanical brake. (True/False) _____

5. The direction of rotation of a DC motor depends on the direction of the magnetic field and has nothing to do with the direction of current flow through the armature. (True/False) _____

6. The use of DC motors is more widespread than the use of AC motors. (True/False) _____

7. Motor starter terminal markings are used to tag terminals to which connections can be made from external circuits. (True/False) _____

8. The rotating part of an AC motor is referred to as the _____; the stationary part is called the _____.

9. A squirrel-cage rotor has no physical electrical connection to it. (True/False) _____

10. The device used to automatically disconnect the starting winding of a single-phase split-phase motor is called a _____ switch.

11. In a dual-voltage split-phase motor, the two run windings are connected in _____ for the high-voltage operation.
   a. Wye
   b. Delta
   c. Parallel
   d. Series
   Answer _____

12. When a dual-voltage motor is operated at the lower voltage rating, the
   a. Operating power is lower
   b. Operating power is higher
   c. Operating current is lower
   d. Operating current is higher
   Answer _____

13. When a split-phase motor is operated with a capacitor in series with one of the stator windings, the
   a. Starting torque is lower
   b. Starting torque is higher
   c. Starting voltage is lower
   d. Starting voltage is higher
   Answer _____

14. Single-phase AC motors are typically constructed in large horsepower sizes. (True/False) _____

15. Three-phase motors are not self-starting on their running windings. (True/False) _____

16. The three distinct windings of a three-phase motor are referred to as

    a. L1, L2, L3

    b. Phase A, phase B, phase C

    c. Phase 1, phase 2, phase 3

    d. T1, T2, T3

    Answer _____

17. All three-phase motors are wired so that the phases are connected in either

    _____ or _____ configuration.

18. On the nine-lead, three-phase, dual-voltage motor connection diagram shown in

    Figure 2-5:

    a. Mark each motor lead (T1, T2, etc.).

    b. Identify each phase coil (A, B, C).

    c. Show the motor properly connected for high-voltage operation.

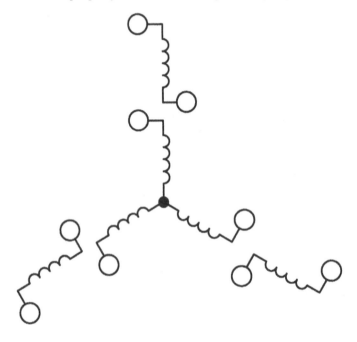

Figure 2-5   Dual-voltage motor diagram for Question 18.

19. The speed of an AC induction motor depends on the

    a. Frequency of the electrical power supply

    b. Voltage of the electrical power supply

    c. Number of poles built into the motor

    d. Both a and c

    Answer _____

20. You can reverse the direction of rotation of a three-phase motor by reversing any two of the three main power leads to the motor. (True/False) _____

21. In a variable-frequency motor drive, the higher the frequency of the power supplied to the motor, the slower is its operating rpm. (True/False) _____

22. Inverter-duty describes a class of motors that are capable of operation

    a. From a variable-frequency drive

    b. From a DC power supply

    c. For intermittent duty only

    d. For continuous duty

    Answer _____

PART 4   Motor Nameplate and Terminology

1. Information required to identify a motor is contained on the motor _____.

2. NEMA requires that a motor be able to carry its rated horsepower at nameplate voltage plus or minus

    a. 2 percent

    b. 5 percent

    c. 8 percent

    d. 10 percent

    Answer _____

3. Motors that are not fully loaded draw less current than their rated nameplate current. (True/False) _____

4. The line frequency of a motor is abbreviated on the nameplate as

    a. CY

    b. CYC

    c. Hz

    d. All of these

    Answer _____

5. The phase rating of a motor is listed on the nameplate as _____ , _____ , or _____ .

6. The rated nameplate speed of a motor is its approximate speed with no load applied. (True/False) _____

7. The motor nameplate _____ temperature rating is the maximum rated temperature of the air surrounding the motor.

8. The amount the motor winding temperature will increase above the ambient temperature at full load is specified according to the nameplate

   _____ _____ rating.

9. Thermal images of electric motors reveal their

   a. Internal conductor temperature

   b. Exterior surface temperature

   c. Operating current

   d. All of these

   Answer _____

10. Standard NEMA motor insulation classes are given by

    a. Number classification (1, 2, 3, etc.)

    b. Alphabetic classification (A, B, F, H, etc.)

    c. Color classification (red, blue, green, etc.)

    d. Temperature level (high, medium, low, etc.)

    Answer _____

11. The duty cycle of a motor is listed on the nameplate as _____ or

    _____.

12. The horsepower nameplate rating of a motor is a measure of the full-load output power the shaft of the motor can produce without stalling. (True/False)

    _____

13. The motor nameplate code letter specifies the _____ rotor-current rating of the motor.

14. The motor nameplate design letter is an indication of the motor's torque-speed curve. (True/False) _____

15. A motor nameplate indicates that the motor has a service factor of 1.25. This means that the

    a. Normal routine service time interval can be increased by 125 percent

    b. Cost of operating the motor is 125 percent higher than an equivalent motor rated for a service factor of 1

    c. Motor can on occasion safely develop 125 percent of its rated horsepower

    d. Motor can safely be operated at 125 percent of its rated voltage

    Answer _____

16. The two general classifications of motor enclosures are _____ and

_____ _____.

17. When replacing a motor, selecting the same _____ ensures that the

mounting mechanism and hole positions will match.

    a. Code letter

    b. Design letter

    c. Horsepower

    d. Frame size

    Answer _____

18. The nameplate efficiency value of a motor is its efficiency rating with no load

applied to the motor. (True/False) _____

19. A motor with a low power factor rating will draw less current for the same

horsepower than a motor with a high power factor. (True/False) _____

20. The nameplate of a motor indicates that it is thermally protected. This means

    a. The motor has a built-in thermal device

    b. The motor must be protected by an external thermal device

    c. The motor uses thermal protection instead of a fuse or circuit breaker

    d. The motor is protected against low thermal temperatures

    Answer _____

21. Connect the motor shown in Figure 2-6 for clockwise rotation and low voltage.

VOLTAGE	ROTATION	L1	L2	JOIN
HIGH	CCW	1	4, 5	2, 3, 8
	CW	1	4, 8	2, 3, 5
LOW	CCW	1, 3, 8	2, 4, 5	————
	CW	1, 3, 5	2, 4, 8	————

Figure 2-6   Motor connection diagram for Question 21.

22. The *plugging* of a motor refers to

    a. Braking by reverse rotation

    b. Momentary operation or small movement of a machine

    c. Automatic restarting of a motor after a power failure

    d. Applying a reduced supply of voltage to a motor during starting

    Answer _____

23. The motor term *auxiliary contact* refers to

    a. A contact that is operated electromechanically

    b. A contact that is operated from some remote point

    c. The contact of a switching device in addition to the main circuit contact

    d. A contact that provides running overcurrent protection

    Answer _____

24. With *low-voltage protection* motor control

    a. A power failure disconnects service and, when power is restored, manual restarting is required

    b. A power failure disconnects service and, when power is restored, the controller automatically restarts

    c. Only two wires are required for the control device

    d. Both b and c

    Answer _____

25. A *motor starter* is an electric controller used to _____ a connected motor

    a. Start

    b. Stop

    c. Protect

    d. All of these

    Answer _____

PART 5   Manual and Magnetic Motor Starters

1. Which of the following types of motor starters connects the motor directly to the supply line on starting?

    a. Start/stop starter

    b. Jogging starter

    c. Full-voltage starter

    d. Reduced-voltage starter

    Answer _____

2. With a selector switch, a

    a. Plunger actuates the contacts

    b. Button actuates the contacts

    c. Relay actuates the contacts

    d. Rotating handle actuates the contacts

    Answer _____

3. Manual motor starters allow a motor to be controlled from any location.

    (True/False) _____

Answer Questions 4 to 8 with reference to the motor control circuit shown in Figure 2-7.

Figure 2-7   Motor control diagram for Questions 4–8.

4. Which is the normally closed push button?

    a. Start button

    b. Jog button

    c. Stop button

    d. Plug button

    Answer _____

5. Which is the normally open contact?

   a. The holding contact

   b. The overload contact

   c. The start contact

   d. The stop contact

   Answer _____

6. Where would you connect another start button?

   a. In series with the stop button

   b. In series with the start button

   c. In parallel with the stop button

   d. In parallel with the start button

   Answer _____

7. Where would you connect another stop button?

   a. In series with the stop button

   b. In series with the start button

   c. In parallel with the stop button

   d. In parallel with the start button

   Answer _____

8. This control circuit would be classified as:

   a. DC control

   b. AC control

   c. Two-wire control

   d. Three-wire control

   Answer _____

9. Two-wire control circuits use a _____-contact type of control device.

10. Three-wire control circuits are designed to protect against automatic

    _____ when power returns after a power failure.

11. In the circuit in Figure 2-8, pressing the push button will:

    b. Switch both lights on

    c. Switch both lights off

    d. Switch light A on and light B off

    e. Switch light A off and light B on

    Answer _____

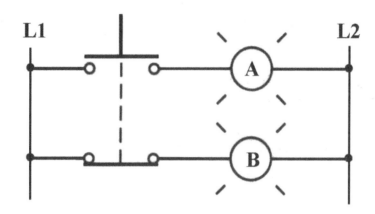

Figure 2-8   Control circuit for Question 11.

12. In the circuit in Figure 2-9, the light is switched on:

    a. Only when PB1 is pressed

    b. Only when PB2 is pressed

    c. When either PB1 or PB2 is pressed

    d. When both PB1 and PB2 are pressed

    Answer _____

Figure 2-9   Control circuit for Question 12.

13. In the circuit in Figure 2-10, which circuit state will switch the light on?

   a. PB1 and PB2 pressed—PB3 not pressed

   b. PB1 and PB3 pressed—PB2 not pressed

   c. PB2 and PB3 pressed—PB1 not pressed

   d. PB2 pressed—PB1 and PB3 not pressed

   Answer _____

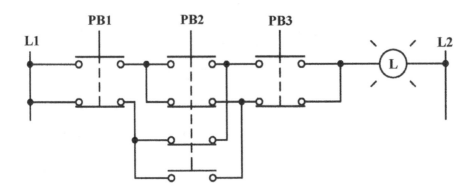

Figure 2-10   Control circuit for Question 13.

hands on PRACTICAL ASSIGNMENTS

1. The purpose of this assignment is to develop a ladder diagram from a wiring diagram. For the magnetic motor starter shown in Figure 2-11, redraw the control circuit in the form of a ladder diagram.

Figure 2-11   Magnetic motor starter for Practical Assignment 1.

   Answer

2. The purpose of this assignment is to properly connect and perform measurements on a single-phase, dual-voltage motor installation. Complete the following tasks using whatever single-phase, dual-voltage motor is available to you:

- Connect the motor for a given voltage.
- Reverse the direction of rotation of the motor.
- Measure the motor current.
- Measure the motor speed.

3. The purpose of this assignment is to properly connect and perform measurements on a three-phase, dual-voltage motor installation. Complete the following tasks using whatever three-phase, dual-voltage motor is available to you:

- Connect the motor for a given voltage.
- Reverse the direction of rotation of the motor.
- Measure the motor current.
- Measure the motor speed.

4. The purpose of this assignment is to connect a three-phase manual motor starter.

- Complete a wiring of the circuit shown in Figure 2-12.
- Wire the circuit in a neat and professional manner.

Figure 2-12   Manual motor starter for Practical Assignment 4.

5. a. The purpose of this assignment is to connect a magnetic across-the-line motor starter to operate a three-phase motor using a pilot device connected for two-wire control. Complete the wiring diagram for the two-wire control circuit shown in Figure 2-13. Wire the circuit in a neat and professional manner.

    b. Does the circuit provide low-voltage release or low-voltage protection?

    _____

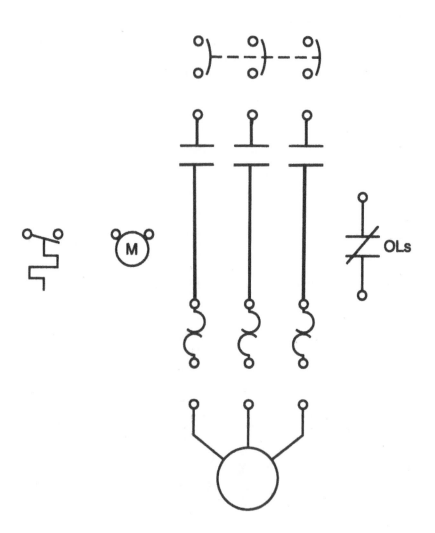

Figure 2-13 Two-wire control circuit for Practical Assignment 5.

6. a. The purpose of this assignment is to connect a magnetic across-the-line motor starter to operate a three-phase motor using a start and a stop push button connected for three-wire control. Complete the wiring diagram for the three-wire control circuit shown in Figure 2-14. Wire the circuit in a neat and professional manner.

   b. Does the circuit provide low-voltage release or low-voltage protection? _____

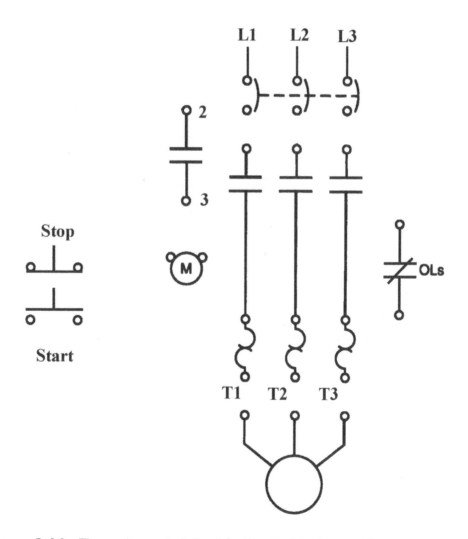

Figure 2-14 Three-wire control circuit for Practical Assignment 6.

"THE CONSTRUCTOR" CIRCUIT ANALYSIS ASSIGNMENTS

1. The purpose of this assignment is to observe the operation of a two-wire motor control circuit. Download circuit file Chapter 2 Simulation Assignment 01 (Figure 2-15, p. 27) from the assignment diagrams folder. Click on the Main Power Switch (ON/OFF) located on the left side of the toolbar. Left-click on the circuit breaker to close it. Left-click on the temperature switch to alternate between closed and open states.

    a. Outline what happens when the temperature switch is closed.

    _____

    _____

    b. Outline what happens when the temperature switch is opened.

    _____

    c. With the temperature switch closed, open and close the circuit breaker to remove and apply power to the circuit. Outline the reaction of the motor when the power is removed and reapplied.

    _____

    _____

    d. With the motor operating, press shift and left-click on the motor to simulate a motor overload condition. Outline how the control circuit reacts to this overload condition.

    _____

    _____

Figure 2-15   Two-wire control circuit.

2. The purpose of this assignment is to observe the operation of a three-wire motor control circuit. Download circuit file Chapter 2 Simulation Assignment 02 (Figure 2-16, p. 28) from the assignment diagrams folder. Click on the Main Power Switch (ON/OFF) and left-click on the main switch to close it. Left-click on the start and stop push buttons to operate them.

   a. Outline what happens when the start button is momentarily closed.

        _____

        _____

   b. With the motor operating, outline what happens when the stop button is momentarily opened.

        _____

        _____

   c. With the motor operating, open and close the main switch to remove and apply power to the circuit. Outline the reaction of the motor when the power is removed and reapplied.

        _____

        _____

d. With the motor operating, press shift and left-click on the motor to simulate a motor overload condition. Outline how the motor circuit reacts to this overload condition.

_____

_____

Figure 2-16 Three-wire control circuit.

3 Motor Transformers and Distribution Systems

PART 1   Power Distribution Systems

Place the answers in the space provided.

1. The _____ station of power generation and distribution enables power to be produced at one location for immediate use at other distant locations.

2. Without the use of _____ to efficiently raise and lower voltage levels, the widespread distribution of electrical power would be impractical.

3. High voltages are used in transmission lines to reduce the transmission

    _____.

    a. Resistance
    b. Current
    c. Power
    d. Power factor
    Answer _____

4. Large diameter conductors are used in electrical utility transmission systems relative to the power transmitted. (True/False) _____

5. The higher the level of the distribution voltage, the more difficult and expensive it becomes to provide for adequate insulation of the conductors. (True/False) _____

6. Three-phase power is usually supplied to commercial and industrial customers. (True/False) _____

7. For the distribution system shown in Figure 3-1, determine the value of the transmission current. Answer _____

Figure 3-1   Circuit diagram for Question 7

8. Compared to the low-voltage side, the high-voltage side of a power transformer draws _____ current.

    a. More

    b. Less

    c. The same

    d. A fixed amount of

    Answer _____

9. Substations contain _____ transformers that reduce the transmission voltage levels.

10. The three major sections of the unit substation are

    a. _____

    b. _____

    c. _____

11. The unit substation is left open on the back for convenient access to all parts. (True/False) _____

12. Before attempting to open the primary switch on a unit substation, you should disconnect the loads from the transformer. (True/False) _____

13. Electrical distribution systems within buildings are required to safely deliver electrical energy without any component overheating or unacceptable voltage drops. (True/False) _____

14. Identify the three major sections of the electrical distribution system shown in Figure 3-2.

   a. _____

   b. _____

   c. _____

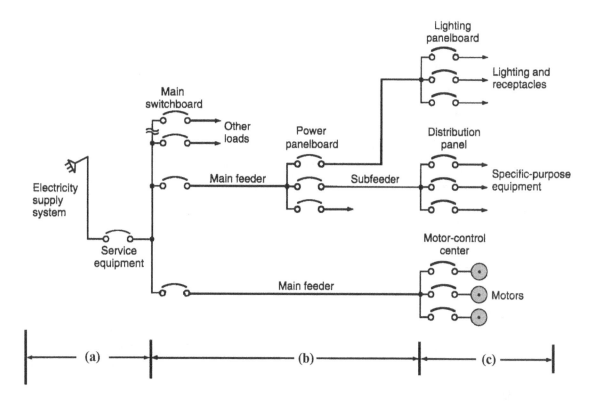

Figure 3-2   Electrical distribution system for Question 14.

15. Correct selection of conductors for feeders and branch circuits must take into account _____, _____, and _____ requirements.

16. In installing motors and motor controllers, which of the following needs to be adhered to?

   a. NEC

   b. State and local codes

   c. Manufacturers' instructions

   d. All of these

   Answer _____

17. The unit of electric energy generated by the power station always matches with that distributed to customers. (True/False) _____

18. A power line of resistance R causes a power loss of

    a. $V \times I$

    b. $V^2 \times R$

    c. $I^2 \times R$

    d. $I \times R$

    Answer _____

19. Power line losses due to poor power factor can be reduced by the application of _____ into the system.

20. Identify the three types of raceways shown in Figure 3-3.

    a. _____

    b. _____

    c. _____

(a)   (b)   (c)

Figure 3-3   Raceways for Question 20.

21. Conduit capacity is generally based on a _____ fill ratio.

    a. 90 percent

    b. 80 percent

    c. 60 percent

    d. 40 percent

    Answer _____

22. A panelboard is usually supplied from a switchboard and further divides the power distribution system into smaller parts. (True/False) _____

23. What is the value of the line-to-neutral voltage for the three-phase four-wire panelboard feeder circuit shown in Figure 3-4?

Answer _____

Figure 3-4   Feeder circuit for Question 23.

24. The proper grounding and bonding of an electrical distribution system ensures that any person who comes in contact with any _____ parts of the installation will not receive an electric shock.

25. The impedance of the ground path is kept to a maximum. (True/False) _____

26. The equipment grounding bus is noninsulated and connects directly to the panelboard metal enclosure. (True/False) _____

27. Correctly identify the three-phase buses (A-B-C) for the panel bus arrangement shown in Figure 3-5.

(i) _____ (ii) _____ (iii) _____

Figure 3-5   Panel bus arrangement for Question 27.

28. Main lug type panelboards contain a main breaker that is an integral part of the panelboard. (True/False) _____

29. A motor control center is normally designed to accommodate plug-in–type motor control units. (True/False) _____

PART 2   Transformer Principles

1. Identify the components of the transformer circuit shown in Figure 3-6.

   a. _____   b. _____

   c. _____   d. _____

   e. _____   f. _____

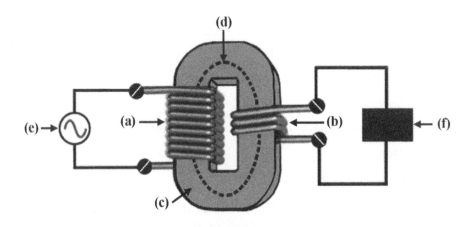

Figure 3-6   Transformer circuit for Question 1.

2. The principle of operation of a transformer is based on _____ induction.

   a. Mutual

   b. Self

   c. Primary

   d. Secondary

   Answer _____

3. Movement of coils within a transformer produces the transformation of voltage. (True/False) _____

4. For an ideal transformer, the power input is equal to the power output.
   (True/False) _____

5. Transformer power is rated in volt-amperes instead of watts.
   (True/False) _____

6. The voltage ratio of a transformer is equal to its turns ratio.
   (True/False) _____

7. A step-up transformer is one in which the secondary winding current is greater than the primary winding current. (True/False) _____

8. For the transformer circuit shown in Figure 3-7 determine the:

   a. Turns ratio _____

   b. Value of the secondary voltage _____

Figure 3-7   Transformer circuit for Question 8.

9. The efficiency of a transformer is typically in the _____ range.
   a. 90 percent
   b. 80 percent
   c. 60 percent
   d. 40 percent
   Answer _____

10. When the secondary winding of a transformer is disconnected from the load,
    a. Zero current flows in the primary
    b. Very little current flows in the primary
    c. The primary current remains the same as under load
    d. The primary current becomes very high
    Answer _____

11. If the secondary circuit of a transformer becomes overloaded, the primary circuit will not be affected. (True/False) _____

12. A transformer is being designed to increase the voltage from 12 V to 120 V. If the primary requires 400 turns of wire, how many turns are required on the secondary? _____

13. For the transformer circuit shown in Figure 3-8 determine the

   a.   Value of the secondary current flow _____

   b.   Value of the primary current flow _____

   c.   Value of the primary voltage _____

Figure 3-8   Transformer circuit for Question 13.

14. The _____ of a transformer is a measure of the proportion of the applied energy that is transferred to the load.

15. Power losses associated with a transformer are generally relatively small. (True/False) _____

16. The output secondary voltage of a transformer remains constant from no-load to the full-load condition. (True/False) _____

17. Magnetizing inrush current in transformer

   a.   Occurs when the transformer is first energized

   b.   Is typically many times the normal full-load current

   c.   Can result in nuisance tripping of the input overcurrent protection device

   d.   All of these

   Answer _____

PART 3   Transformer Connections and Systems

1. The high-voltage windings on a single-phase transformer are identified as X1 and
    X2. (True/False) _____

2. On a transformer, when the lead marked H1 is instantaneously positive, the lead
    marked X1 will be instantaneously _____.

3. A transformer has subtractive polarity when terminal H1 is adjacent to terminal
    X1. (True/False) _____

4. What is the most popular operating voltage of commercial and industrial motor
    control systems?

    a. 120 V

    b. 208 V

    c. 440 V

    d. 600 V

    Answer _____

5. Correctly connect the dual-voltage control transformer circuits shown in
    Figure 3-9 for the input voltage levels given.

Figure 3-9   Dual-voltage transformer circuits for Question 5.

6. The primary of the three-phase transformer connection shown in Figure 3-10 is connected in a _____ configuration and the secondary is connected in a _____ configuration.

Figure 3-10 Three-phase transformer connection for Question 6.

7. In three-phase transformer systems, the constant _____ is used because the transformer phase windings are 120 electrical degrees apart.

a. 3

b. 2.5

c. 1.73

d. 0.707

Answer _____

8. In a Delta-connected transformer secondary, the phase and line voltages are equal. (True/False) _____

9. For the Wye distribution system shown in Figure 3-11, if the voltage between any two line leads is 208 V, the voltage from any line lead to neutral would be _____ V.

Figure 3-11 Wye distribution system for Question 9.

10. An autotransformer has electrical isolation between the primary and secondary circuits. (True/False) _____

11. An autotransformer motor starter reduces the starting current to the motor by

    a.   Inserting resistance in series with the motor

    b.   Reducing the value of the applied voltage

    c.   Reducing the load on the motor

    d.   Reducing the frequency of the applied voltage

    Answer _____

12. Compared to an equivalent traditional transformer, the autotransformer is smaller in size. (True/False) _____

13. The secondary low-voltage side of a potential instrument transformer is usually wound for

    a.   5 volts

    b.   10 volts

    c.   24 volts

    d.   120 volts

    Answer _____

14. For the circuit of Figure 3-12, transformer X is a _____ transformer, while transformer Y is a _____ transformer.

Figure 3-12 Circuit for Question 14.

15. Instrument transformers provide insulation between the instrument and the high voltage of the power circuit. (True/False) _____

16. A current transformer should always have its secondary shorted when there is current flow in the primary winding but no load connected to the secondary. (True/False) _____

hands on PRACTICAL ASSIGNMENTS

1. a. The purpose of this assignment is to connect a magnetic across-the-line motor starter to operate a three-phase motor using start/stop push buttons and a control transformer. Complete the wiring diagram for the magnetic starter circuit shown in Figure 3-13.

   b. Wire the circuit in a neat and professional manner.

Figure 3-13 Control transformer circuit for Practical Assignment 1.

2. The purpose of this assignment is to develop a wiring diagram for a typical Delta-to-Wye transformer bank connection. Connect the primary transformer lines to the distribution system to form the Delta-connected primary of Figure 3-14. Connect the secondary transformer lines to the distribution system to form a Wye-connected secondary that provides 208-V three phase, 208-V single phase, and 120-V single phase. Connect each motor load to the correct power supply.

Figure 3-14 Three-phase transformer bank circuit for Practical Assignment 2.

41

"THE CONSTRUCTOR" CIRCUIT ANALYSIS ASSIGNMENTS

1. The purpose of this assignment is to observe the operation of a properly grounded control transformer circuit. Download circuit file Chapter 3 Simulation Assignment 01 (Figure 3-15) from the assignment diagrams folder. Click on the Main Power Switch (ON/OFF) located on the left side of the toolbar. Left-click on the circuit breaker to close it and observe the normal operation of the circuit.

    a. How is the motor normally started?

    _____

    b. How is the motor normally stopped?

    _____

    c. According to the control transformer primary connection, would the primary line voltage be 480 or 240 V? Why?

    _____

Figure 3-15 Control transformer circuit.

2. The purpose of this assignment is to observe the operation of a properly grounded control transformer circuit that experiences a ground fault condition. Download circuit file Chapter 3 Simulation Assignment 02 (Figure 3-16, p. 43) from the assignment diagrams folder. Click on the Main Power Switch located on

the left side of the toolbar. Left-click on the circuit breaker to close it and observe the operation of the circuit with a ground fault at the connection between the start and stop push buttons. What happens as soon as the main circuit breaker is closed? Why?

_____

_____

Figure 3-16 Ground-faulted control transformer circuit.

3. The purpose of this assignment is to observe the operation of an *incorrectly grounded* control transformer circuit that experiences a ground fault condition. Download circuit file Chapter 3 Simulation Assignment 03 (Figure 3-17, p. 44) from the assignment diagrams folder. Click on the Main Power Switch located on the left side of the toolbar. Left-click on the circuit breaker to close it and observe the operation of the circuit with the same ground fault and the transformer incorrectly grounded at X1 instead of X2.

   a. What happens when the main circuit breaker is closed and the start button is momentarily clicked closed?

   _____

   _____

b. What happens when the motor is running and the stop button is clicked open? Why?

_____

_____

Figure 3-17 Incorrectly connected transformer ground circuit.

4 Motor Control Devices

PART 1   Manually Operated Switches

Place the answers in the space provided.

1. Pilot-duty devices are normally used to switch horsepower-rated loads. (True/False) _____

2. Which of the following would be classified as a primary control device?

    a. Flow switch

    b. Thermostat

    c. Motor starter

    d. Pressure switch

    Answer _____

3. A push button would normally be considered to be a pilot control device. (True/False) _____

4. A manually operated switch is one that is controlled by _____.

5. The abbreviation SPDT stands for a _____ _____ _____ _____ switch.

6. The type of switch shown in Figure 4-1 is a _____.

Figure 4-1   Switch for Question 6.

7. The DC current rating of a switch would have a lower magnitude than the AC rating. (True/False) _____

8. Identify each type of push button shown in Figure 4-2 and draw the NEMA symbol used to represent it.

(a) Type _____

NEMA Symbol

(b) Type _____

NEMA Symbol

(c) Type _____

NEMA Symbol

Figure 4-2   Push buttons for Question 8.

9. The NC push button _____ the circuit when it is pressed and returns to the _____ position when the button is released.

10. The break-make push button is used for _____ control circuits.

11. Electrical _____ are designed to protect their contents from operating environmental conditions.

12. A push button may consist of one or more _____ blocks, a(n) _____ device, and a(n) _____ plate.

13. _____ head pushbutton operators are used for emergency stop push buttons.

14. A pushbutton assembly can only have a one contact block. (True/False) _____

15. Once actuated, _____ type pushbutton operators are required to be actuated a second time to return the contacts to their normal off state.

16. Emergency stop push buttons must be of the maintained contact type. (True/False) _____

17. Draw the NEMA and IEC symbols used to represent a red pilot light.

NEMA   IEC

18. Push-to-test pilot lights can be energized from two separate signals of different voltage values. (True/False) _____

19. All selector switches operate with spring return to provide momentary contact operation. (True/False) _____

20. Drum switches are used for starting and reversing three-phase squirrel-cage motors. (True/False) _____

PART 2   Mechanically Operated Switches

1. A mechanically operated switch is one that is controlled by hand.

   (True/False) _____

2. Limit switches are designed to be operated:

   a. By the machine operator

   b. When a predetermined pressure is reached

   c. When a predetermined temperature is reached

   d. By contact with an object

   Answer _____

3. Identify each of the switch symbols shown in Figure 4-3.

a) _____   b) _____   c) _____

d) _____   e) _____

Figure 4-3   Switch symbols for Question 3.

4. The type of limit switch shown in Figure 4-4 would be classified as a

   a. Lever type
   b. Push roller type
   c. Wobble-stick type
   d. Fork lever type
   Answer _____

Figure 4-4   Limit switch for Question 4.

5. Micro limit switches require a large amount of pressure to be applied to the operating lever in order to actuate the contacts. (True/False) _____

6. The type of temperature switch shown in Figure 4-5 is

   a. Bimetallic type
   b. Capillary type
   c. Float type
   d. Spring type
   Answer _____

Figure 4-5   Temperature switch for Question 6.

7. All temperature switches use only N.O. (normally open) contacts. (True/False) _____

8. The three categories of pressure switches are _____ pressure, _____ pressure, and _____ pressure.

9. Pressure switches open or close when a preset pressure is reached. (True/False) _____

10. The type of switch shown in Figure 4-6 would be classified as a

    a. Pressure type

    b. Flow type

    c. Float type

    d. Temperature type

    Answer _____

Figure 4-6   Switch for Question 10.

11. A flow switch is used to detect the _____ of air or liquid.

PART 3   Sensors

1. Sensors are devices that are used to _____ and _____ the presence of something.

2. A proximity sensor can detect the presence of an object without physical contact. (True/False) _____

3. An inductive proximity sensor can be actuated by both conductive and nonconductive materials. (True/False) _____

4. Identify each component block of the inductive proximity sensor shown in Figure 4-7.

    a. _____   b. _____

    c. _____   d. _____

Figure 4-7   Inductive proximity sensor for Question 4.

5. Draw acceptable NEMA and IEC symbols used to represent an N.O. proximity switch.

NEMA   IEC

6. The proximity sensor shown in Figure 4-8 would be classified as a

   a. Three-wire type

   b. Two-wire type

   c. DC type

   d. Both a and c

   Answer _____

Figure 4-8   Proximity sensor for Question 6.

7. A proximity sensor should not normally be used to directly operate a motor. (True/False) _____

8. The electronic switching circuit of a sensor:

   a. Has infinite resistance in the off state

   b. Will allow a small amount of current to flow through it in the off state

   c. Will have a small voltage drop across it in the on state

   d. Both b and c

9. When a metal object enters the high-frequency field of an inductive proximity sensor, it causes

   a. A change in amplitude of the oscillating circuit

   b. The oscillator circuit to start oscillating

   c. Vibration of the oscillator

   d. Both b and c

   Answer _____

10. In order to operate properly, a two-wire proximity sensor should be powered continuously. (True/False) _____

11. The area between the operating and release points of a proximity sensor is known as the _____ zone.

12. The position identified on the field of the proximity sensor shown in Figure 4-9 is known as the

    a. Saturation point

    b. Hysteresis point

    c. Operating point

    d. Release point

    Answer _____

Figure 4-9   Proximity sensor for Question 12.

13. A capacitive proximity sensor produces a(n)

    a. Electromagnetic sensing field

    b. Electrostatic sensing field

    c. Constant current sensing field

    d. Constant voltage sensing field

    Answer _____

14. When the target enters the field of a capacitive proximity sensor, the oscillator circuit begins oscillating. (True/False) _____

15. Capacitive proximity sensors will sense only nonmetallic materials. (True/False) _____

16. Identify the two basic components of the photoelectric sensor shown in Figure 4-10.

    a. _____

    b. _____

Figure 4-10   Photoelectric sensor for Question 16.

17. Identify the type of scan technique used for each of the photoelectric sensors shown in Figure 4-11.

(a) _____   (b) _____   (c) _____

Figure 4-11   Photoelectric sensors for Question 17.

18. Which of the following scan techniques is best suited for long-range scanning?
    a. Through-beam
    b. Retroreflective
    c. Polarized retroreflective
    d. Diffuse
    Answer _____

19. A _____ retroreflective scan sensor is designed to be able to detect shiny objects.

20. Fiber optics can be used with through-beam, retroreflective, or diffuse scan sensors. (True/False) _____

21. Fiber optic sensor systems are completely immune to all forms of electrical interference. (True/False) _____

22. Hall-effect sensors are used detect both the proximity and the _____ of a magnetic field.

23. Identify the measurement application for the Hall-effect sensing circuits shown in Figure 4-12.

a. Application:

_____

_____

_____

b. Application:

_____

_____

_____

Figure 4-12   Hall-effect sensors for Question 23.

24. An ultrasonic sensor operates by sending high-frequency _____ waves toward the target and measuring the _____ it takes for the pulses to bounce back.

25. For the ultrasonic sensor shown in Figure 4-13, the sensor will generate a signal of approximately _____ mA when the tank is full and _____ mA when the tank is empty.

Figure 4-13   Ultrasonic sensor for Question 25.

26. Identify each type of temperature sensor shown in Figure 4-14.

(a) _____ (b) _____ (c) _____ (d) _____

Figure 4-14 Temperature sensors for Question 26.

27. Which of the following temperature sensors operates on the principle that electrical resistance of metals is directly proportional to temperature?
    a. IC
    b. RTD
    c. Thermocouple
    d. Thermistor
    Answer _____

28. The output voltage of a thermocouple is proportional to the difference in voltage between the _____ junction and the _____ junction.

29. Different thermocouple types basically have the same voltage output curves. (True/False) _____

30. Which of the following temperature sensors provides the most accurate measurement of temperature?
    a. IC
    b. RTD
    c. Thermocouple
    d. Thermistor
    Answer _____

31. The material most often used for resistance temperature detectors is
    a. Platinum
    b. Gold
    c. Copper
    d. Aluminum
    Answer _____

32. As the temperature of a thermistor with a negative temperature coefficient increases, its resistance decreases. (True/False) _____

33. The operating temperature range of the IC temperature sensor is much lower than that of a thermocouple. (True/False) _____

34. _____ are tubular fittings used to protect temperature sensors installed in industrial processes.

    Answer _____

35. A tachometer generates a voltage that is proportional to _____.

36. Identify the parts of the magnetic pickup sensor shown in Figure 4-15.

    a. _____

    b. _____

    c. _____

    d. _____

Figure 4-15   Magnetic pickup for Question 36.

37. A magnetic pickup sensor generates pulses that are counted to determine shaft speed. (True/False) _____

38. An encoder sensor is used to convert linear or rotary motion into a binary _____ signal.

39. Encoder sensors are used in applications such as robotic control where _____ have to be accurately determined.

40. Identify each type of flowmeter shown in Figure 4-16.

    a. _____

    b. _____

Figure 4-16   Flowmeters for Question 40.

41. The usual approach taken to measure fluid flow is to convert the _____ energy that the fluid has into some other measurable form.

42. Which of the flowmeters operates by measuring the changes of induced voltage of a conductive fluid?

    a. Magnetic

    b. Target

    c. Turbine

    d. Differential pressure

    Answer _____

PART 4   Actuators

1. An actuator is any device that converts an electrical signal into

   _____ _____.

2. The contacts of an electromechanical relay are switched by energizing and de-energizing a(n) _____.

3. The normally closed contact of a relay is closed any time the coil is energized. (True/False) _____.

4. Compared to a single-break contact, a double-break relay contact

   a. Uses two pair of contacts

   b. Can handle higher voltages

   c. Provides longer contact life

   d. All of these

   Answer _____

5. A dry contact usually means that it is completely isolated and has no voltage applied to it from its initiating equipment. (True/False) _____

6. A solenoid is made up of coil with a movable _____.

7. Which type of solenoid is shown in Figure 4-17?

   a. AC

   b. DC

   c. Rotary

   d. Manual

   Answer _____

Figure 4-17 Solenoid for Question 7.

8. An AC solenoid rated for 120-V alternating current, if connected to a 120-V direct current source, would draw a much lower than normal current. (True/False) _____.

9. If a DC solenoid sticks in the open position, a burnout of the coil is likely. (True/False) _____.

10. In AC solenoids, coil current decreases as the plunger moves toward the seated position. (True/False) _____.

11. A solenoid valve is a combination of a solenoid coil operator and

    a. Contacts

    b. Sensor

    c. Transmitter

    d. Valve

    Answer _____

12. When a solenoid valve is electrically energized it can be designed to

    a. Open the flow of media

    b. Shut off the flow of media

    c. Direct the flow of media

    d. All of these

    Answer _____

13. Stepper motors rotate continuously when voltage is applied to their terminals. (True/False) _____.

14. For a DC stepper motor, the amount of rotation is directly proportional to the _____ of pulses and the speed to the _____ of the pulses.

15. Generally, stepper motors are rated for more than 1 hp and are used to operate very heavy loads. (True/False) _____.

16. Stepper motor systems can provide precise position control of movement without the use of a feedback signal. (True/False) _____.

17. An open-loop motor control system is one that is operated

    a. Without a feedback signal

    b. With a feedback signal

    c. Without a power signal

    d. With a power signal

    Answer _____

18. All servo motors operate using a closed-loop control system.

    (True/False) _____.

19. Identify all blocks for the servo motor control system shown in Figure 4-18.

    a. _____

    b. _____

    c. _____

    d. _____

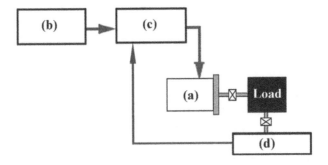

Figure 4-18　Servo system for Question 19.

20. Brushless DC motors are mechanically commutated.

    (True/False) _____.

hands on   PRACTICAL ASSIGNMENTS

1. a. The purpose of this assignment is to connect pilot lights to a magnetic across-the-line motor starter circuit. Pilot lights are to be connected so that the green light is on when the starter coil is deenergized and the red light is on when the starter coil is energized. Complete the wiring diagram for the magnetic across-the-line motor starter circuit of Figure 4-19.

   b. Wire the circuit in a neat and professional manner.

Figure 4-19   Magnetic across-the-line motor starter circuit for Practical Assignment 1.

"THE CONSTRUCTOR" CIRCUIT ANALYSIS ASSIGNMENTS

1. The purpose of this assignment is to observe the operation of a break-make pushbutton circuit. Download circuit file Chapter 4 Simulation Assignment 01 (Figure 4-20) from the assignment diagrams folder. Click on the Main Power Switch (ON/OFF) located on the left side of the toolbar. Left-click on the circuit breaker to close it. Operate the break-make push button by left-clicking on it.

    a. What is the state of the pilot lights when the button is not actuated?

    _____

    b. What is the state of the pilot lights when the button is actuated?

    _____

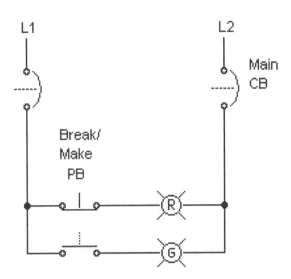

Figure 4-20   Break-make pushbutton circuit.

2. The purpose of this assignment is to observe the operation of an emergency stop pushbutton circuit. Download circuit file Chapter 4 Simulation Assignment 02 (Figure 4-21) from the assignment diagrams folder. Click on the Main Power Switch (ON/OFF) located on the left side of the toolbar. Left-click on the circuit breaker to close it. Operate the circuit control push buttons and observe what happens.

   a. What type of contacts (maintained or momentary) are used for the start and stop push buttons?

   _____

   b. With the motor running, actuate the E-Stop button. How is the motor restarted once the E-Stop button is pressed?

   _____

   _____

   c. What type of contact (maintained or momentary) is used for the emergency stop push button?

   _____

Figure 4-21 Emergency stop pushbutton circuit.

3. The purpose of this assignment is to observe the operation of pilot lights used in conjunction with a magnetic motor starter circuit. Download circuit file Chapter 4 Simulation Assignment 03 (Figure 4-22) from the assignment diagrams folder. Click on the Main Power Switch (ON/OFF) located on the left side of the toolbar. Left-click on the main breaker to close it.

a. What is the initial status of the pilot lights?

_____

b. Start the motor. What is the new status of the pilot lights?

_____

c. Right-click on the NC auxiliary M contact and lock it in the open position. What effect does this has on the operation of the pilot lights?

_____

_____

Figure 4-22 Motor starter pilot light circuit.

4. The purpose of this assignment is to observe the operation of a push-to-test pilot light circuit. Download circuit file Chapter 4 Simulation Assignment 04 (Figure 4-23) from the assignment diagrams folder. Click on the Main Power Switch (ON/OFF) located on the left side of the toolbar. Left-click on the main breaker to close it.

a. What is the pilot light connected to indicate?

_____

b. Start the motor. What is the status of the pilot light?

_____

c. With the motor running, left-click the pilot light. What is the status of the pilot light?

_____

d. Stop the motor. What is the status of the pilot light?

_____

e. With the motor stopped, left-click the pilot light. What is the status of the pilot light?

_____

Figure 4-23   Push-to-test pilot light circuit.

5. The purpose of this assignment is to observe the operation of a three-position selector switch circuit. Download circuit file Chapter 4 Simulation Assignment 05 (Figure 4-24) from the assignment diagrams folder. Click on the Main Power Switch (ON/OFF) located on the left side of the toolbar. Left-click on the main breaker to close it.

   a. With the selector switch in the manual (M) position, what device controls the operation of the motor?

   _____

   b. Right-click the selector switch to the OFF position. Why will the motor not operate with the switch in this position?

   _____

   c. Right-click the selector switch to the auto (A) position. What device controls the operation of the motor with the switch in this position?

   _____

Figure 4-24 Three-position selector switch circuit.

6. The purpose of this assignment is to observe the operation of a two-pole limit switch circuit. Download circuit file Chapter 4 Simulation Assignment 06 (Figure 4-25) from the assignment diagrams folder. Click on the Main Power Switch (ON/OFF) located on the left side of the toolbar. Left-click on the main breaker to close it.

   a. What is the initial state of the pilot lights?

   _____

   b. Left-click the limit switch icon to activate the contacts. What is the state of the pilot lights when the limit switch is actuated?

   _____

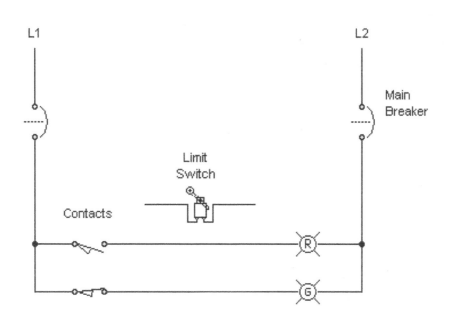

Figure 4-25 Two-pole limit switch circuit.

7. The purpose of this assignment is to observe the operation of a limit switch circuit used to provide overtravel protection. Download circuit file Chapter 4 Simulation Assignment 07 (Figure 4-26) from the assignment diagrams folder. Click on the Main Power Switch (ON/OFF) located on the left side of the toolbar. Left-click on the main breaker to close it.

   a. Left-click on the forward button to operate the motor in the forward direction. While the motor is operating in the forward direction, left-click on the reverse button. What prevents the reverse coil from energizing?

   _____

   b. While the motor is operating in the forward direction, actuate the forward overtravel limit switch. Why can only the reverse coil now be energized?

   _____

   _____

   c. Assume the reverse overtravel limit switch is actuated and fails to return to its normal state once travel in the forward direction has been initiated. How would this affect the operation of the circuit?

   _____

   _____

Figure 4-26   Limit switch overtravel protection.

8. The purpose of this assignment is to observe the operation of a temperature switch control of a motor circuit. Download circuit file Chapter 4 Simulation Assignment 08 (Figure 4-27) from the assignment diagrams folder. Click on the Main Power Switch (ON/OFF) located on the left side of the toolbar. Left-click on the manual starter to close it.

   a. Left-click the selector switch to the hand (H) position. Which selector switch contacts are closed in this position?

   _____

   b. Right-click the selector switch to the auto (A) position. Which selector switch contacts are closed in this position?

   _____

Figure 4-27   Temperature switch control of a motor circuit.

9. The purpose of this assignment is to observe the operation of an interposing relay in a motor control circuit. Download circuit file Chapter 4 Simulation Assignment 09 (Figure 4-28) from the assignment diagrams folder. Click on the Main Power Switch (ON/OFF) located on the left side of the toolbar. Left-click on the main breaker and on/off switch to close them.

a. Left-click on the proximity switch to close it and outline the sequence of events that occur in order to start the motor.

_____

_____

b. What is the operating voltage of the proximity switch? _____

c. What is the operating voltage of coil CR? _____

d. What is the operating voltage of contact CR? _____

e. What is the operating voltage of coil M? _____

Figure 4-28   Interposing relay motor control circuit.

10. The purpose of this assignment is to observe the operation of solenoid valves in a tank filling and emptying circuit. Download circuit file Chapter 4 Simulation Assignment 10 (Figure 4-29) from the assignment diagrams folder. Click on the Main Power Switch (ON/OFF) located on the left side of the toolbar.

a. Left-click the fill button to initiate filling of the tank. Left-click on the full tank sensor to simulate a full tank. List the sequence of operation that takes place to automatically fill the tank.

_____

_____

_____

b. Once the tank is full, left-click the empty button and list the sequence of operations that takes place to automatically empty the tank.

_____

_____

_____

Figure 4-29 Operation of solenoid valves in a tank filling and emptying circuit.

5 Electric Motors

PART 1   Motor Principle

Place the answers in the space provided.

1. Electric motors are used to convert _____ energy into

   _____ energy.

2. An electric motor uses _____ and electric currents to operate.

3. The two basic categories of motors are _____ and

   _____.

4. The magnetic field of a magnet is represented by

   a. Positive charges

   b. Negative charges

   c. Lines of current

   d. Lines of flux

   Answer _____

5. Draw the shape and indicate the direction of the magnetic field for the conductor

   shown in Figure 5-1.

Figure 5-1   Circuit for Question 5.

6. The magnetic field produced by a current-carrying coil resembles that of a

   permanent magnet. (True/False) _____

7. The polarity of the poles of a current-carrying coil reverses whenever the current

   flow through the coil reverses. (True/False) _____

8. An electric generator is a machine that uses magnetism to convert _____ energy into _____ energy.

9. AC generators are also called _____ because they produce an alternating current.

    a. Alternators

    b. Commutators

    c. Motors

    d. Rectifiers

10. A generator produces electricity by

    a. Aligning magnetic lines of force

    b. Increasing voltage

    c. Rotating a loop of wire in a magnetic field

    d. Decreasing resistance

11. The commutator of a DC generator converts the generated AC voltage into DC voltage. (True/False) _____

12. An electric motor rotates as the result of the interaction of two magnetic fields. (True/False) _____

13. Complete the drawing of Figure 5-2 showing

    • The direction of the electron flow through the conductor

    • The direction of the magnetic field of the permanent magnets

    • The direction in which the conductor would move

Figure 5-2   Circuit for Question 13.

PART 2   Direct Current Motors

1. Direct current motors are used where a wide range of high torque and speed control is required. (True/False) _____

2. Use of DC motors is limited by the fact that

   a. Electric utility systems deliver alternating current

   b. DC motor types are more expensive to construct than AC types

   c. DC motor types require more maintenance than AC types

   d. All of these

   Answer _____

3. Identify the major components of the DC motor shown in Figure 5-3.

   a. _____

   b. _____

   c. _____

   d. _____

   e. _____

   f. _____

   g. _____

   h. _____

Figure 5-3   Motor for Question 3.

4. The rotational speed of a motor's shaft is usually measured in _____.

5. _____ refers to the turning force supplied by the motor's shaft.

6. One horsepower is the equivalent of _____ watts of electrical power.

7. The action of switching current between coils within the armature of a DC motor is called _____.

8. DC permanent-magnet motors are often used in servo motor applications. (True/False) _____

9. Answer each of the following with reference to the DC motor shown in Figure 5-4.

   a. This motor would be classified as a _____ type.

   b. The direction of rotation is changed by _____

   _____

   c. The speed is varied by _____

   _____

Figure 5-4   Motor circuit for Question 9.

10. A series-type DC motor uses field coils with

    a. Low resistance and made up of a few turns of large-diameter wire

    b. Low resistance and made up of a many turns of small-diameter wire

    c. High resistance and made up of a few turns of large-diameter wire

    d. High resistance and made up of a many turns of small-diameter wire

    Answer _____

11. The series-type DC motor has

    a. Good speed regulation

    b. Low starting current

    c. High starting torque

    d. Both a and c

    Answer _____

12. The no-load speed of a series motor can increase to the point of damaging the motor. (True/False) _____

13. The shunt-type DC motor is known to have

    a. High staring torque

    b. Good speed regulation

    c. Poor voltage regulation

    d. Heavy-gauge field windings

    Answer _____

14. The unmarked armature and shunt field leads for the DC shunt motor shown in Figure 5-5 are to be identified using an ohmmeter. Explain how you would proceed to identify the armature and field leads.

_____

_____

Figure 5-5   DC shunt motor for Question 14.

15. The field winding of a shunt motor is always connected across the armature and cannot be separately excited or connected to a voltage source other than that of the armature. (True/False) _____

16. Label the field and armature leads for the compound generator shown in Figure 5-6.

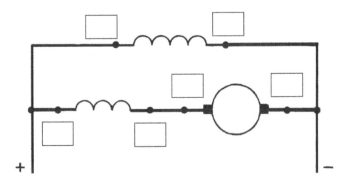

Figure 5-6   Compound generator for Question 16.

17. Identify the type of wound-field DC motor represented by each of the speed characteristic curves shown in Figure 5-7.

   a. _____

   b. _____

   c. _____

Figure 5-7   Characteristic curves for Question 17.

18. A DC motor is to be used to operate the hoist motor of an overhead crane. Which of the following DC motors would be best suited for this application?

   a. Permanent magnet

   b. Series

   c. Shunt

   d. Compound

   Answer _____

19. The speed and torque characteristics shown in Figure 5-8 are that of a

    _____ DC motor.

    a. Permanent magnet

    b. Series

    c. Shunt

    d. Compound

    Answer _____

Figure 5-8   Speed and torque characteristics for Question 19.

20. The direction of rotation of a shunt-wound DC motor is changed by

    a. Reversing the current flow through the armature

    b. Reversing the leads of the DC voltage source

    c. Reversing the current flow through the field

    d. Either a or c

    Answer _____

21. The voltage induced into the armature of a DC motor is known as the

    a. Armature reaction

    b. *IR* voltage drop

    c. Applied voltage

    d. CEMF

    Answer _____

22. The armature current of a 125-V DC motor is measured and found to be 5 A. Determine the amount of counter EMF being generated in the armature if the armature resistance is 1.5 Ω.

    Answer _____

23. At the moment of starting, no CEMF is generated in the armature of a DC motor. (True/False) _____

24. Armature reaction in a DC motor causes

    a. A shift in the neutral plane

    b. Reduction of motor torque

    c. Arcing at the brushes

    d. All of these

    Answer _____

25. Stator poles placed between main field poles, as shown in Figure 5-9, are called _____ and are used to minimize the effects of armature _____.

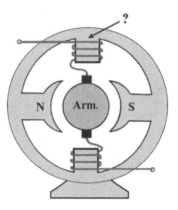

Figure 5-9   Stator poles for Question 25.

26. Determine the full-load speed for a motor with a no-load speed of 2,500 rpm and 2.5 percent speed regulation.

    Answer _____

27. The _____ speed of a DC motor is the speed at which the motor will operate with full rated armature voltage and field current applied.

28. Increasing the voltage applied to the armature of a DC motor will cause it to run slower. (True/False) _____

29. Decreasing the shunt field current of a DC motor will cause it to run slower. (True/False) _____

30. Armature-controlled DC motors are capable of providing rated torque at any speed between zero and the base speed of the motor. (True/False) _____

31. If power to the field circuit of a shunt motor is lost during operation, the motor can accelerate to a dangerously high speed. (True/False) _____

32. Shunt motors are normally started with weakened field current. (True/False) _____

33. Answer each of the following with reference to the DC motor drive block diagram shown in Figure 5-10.

    a. The two basic sections of the drive are the _____ section and the _____ section.

    b. What signal sets the desired speed?

        _____

    c. What might the feedback device be?

        _____

    d. For the block diagram of Figure 5-10, what is the function of the converter block? _____

Figure 5-10 DC motor drive for Question 33.

PART 3   Three-Phase Alternating Current Motors

1. A _____ magnetic field is the key to the operation of AC motors.

2. The synchronous speed of an AC induction motor is always greater than the actual rotor speed. (True/False) _____

3. The higher the frequency of the power source, the _____ (faster/slower) the speed of an AC induction motor.

4. The higher the number of stator poles, the _____ (faster/slower) the speed of an AC induction motor.

5. The synchronous speed of an eight-pole, 60-Hz, AC induction motor would be _____.

6. Figure 5-11 shows a _____-type rotor.

Figure 5-11 Rotor for Question 6.

7. There are no slip rings or any voltage supplied to the rotor of a squirrel-cage induction motor. (True/False) _____

8. The resistance of the squirrel-cage rotor has little effect on the operation of the motor. (True/False) _____

9. When a load is applied to a three-phase squirrel-cage motor the rotor speed will
   a. Increase
   b. Decrease slightly
   c. Remain the same
   d. Fluctuate
   Answer _____

10. For the motor shown in Figure 5-12, when the reverse contacts close, L1, L2, and L3 are connected to _____, _____, and _____ respectively.

11. If one phase to a three-phase motor is lost during operation, the motor will immediately stop on its own. (True/False) _____

12. An induction-type motor needs slip to provide the torque for the motor. (True/False) _____

Figure 5-12
Motor circuit for Question 10.

13. Determine the slip in rpm for a four-pole induction motor with a rotor speed of 1,750 rpm, operating from a 60-Hz supply.

    Answer _____

14. Loading of an induction motor is similar to that of a transformer in that the operation of both involves changing _____ linkages.

15. Power factor (PF) is the ratio of the _____ to the _____ and is a measure of how effectively the current drawn by a motor is converted into useful work.

16. The power factor of an induction motor improves when operating under load. (True/False) _____

17. The locked-rotor starting current of an induction motor is

    a. Equal to the full-load current

    b. Less than the full-load current

    c. One-half the full-load current

    d. Several times the full-load current

    Answer _____

18. _____-pole motors are single-winding motors that will operate at two different speeds, depending on how the windings are connected to form a different number of magnetic poles.

19. The low speed on a separate winding two-speed motor is always one-half of the higher speed. (True/False) _____

20. Connect the dual-speed, six-lead motor shown in Figure 5-13 according to the NEMA nomenclature for high-speed operation.

Figure 5-13

Circuit for Question 20.

NEMA NOMENCLATURE—6 LEADS

CONSTANT TORQUE CONNECTION
Low-speed horsepower is half
of high-speed horsepower.

SPEED	L1	L2	L3		TYPICAL CONNECTION
HIGH	6	4	5	1&2&3 JOIN	2 WYE
LOW	1	2	3	4-5-6 OPEN	1 DELTA

21. Connect the dual-voltage, nine-lead Wye-connected motor shown in Figure 5-14 according to the NEMA nomenclature for low-voltage operation.

Figure 5-14 Wye-connected motor for Question 21.

22. Answer each of the following with reference to the wound-rotor induction motor circuit shown in Figure 5-15.

    a. The motor is normally started with _____ (full/zero) external resistance in the rotor circuit that is gradually reduced to _____ (full/zero).

    b. The motor produces a very _____ (high/low) starting torque from zero speed to full speed at a relatively _____ (high/low) starting current.

    c. The direction of rotation is changed by interchanging any two of the external resistor leads. (True/False) _____

Figure 5-15   Induction motor circuit for Question 22.

23. A three-phase synchronous motor requires

    a. Both AC and DC voltages

    b. Only an AC voltage

    c. Only a DC voltage

    d. Two levels of AC voltage

    Answer _____

24. The three-phase synchronous motor has zero slip. (True/False) _____

25. For the three-phase distribution system shown in Figure 5-16, state whether the system power factor meter is most likely to move in a lead or lag direction for each of the following operating changes:

a. Decrease in the number of inductive motor loads connected to the system
_____

b. Full-load conditions on all inductive-type motor loads connected to the system _____

c. Increase in the synchronous motor rotor field current _____

d. Decrease in the synchronous motor rotor field current _____

e. Connection of a three-phase furnace resistive heater element load to the system _____

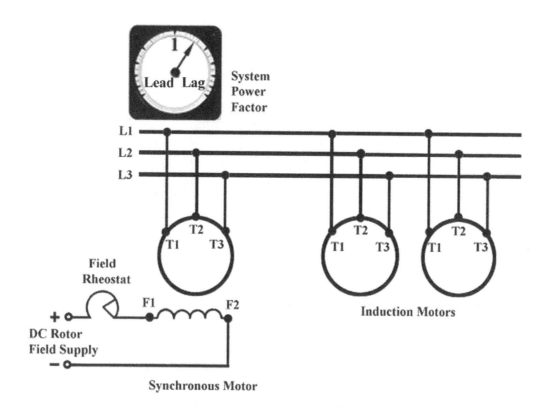

Figure 5-16 Three-phase distribution system for Question 25.

PART 4   Single-Phase Alternating Current Motors

1. Single-phase induction motors are used in residential and commercial applications where _____ power is not available.

2. Unlike three-phase induction motors, single-phase types are not self-starting. (True/False) _____

3. Single-phase motors are classified by their _____ and _____ characteristics.

4. Identify the major components of the split-phase motor shown in Figure 5-17.

    a. _____

    b. _____

    c. _____

    d. _____

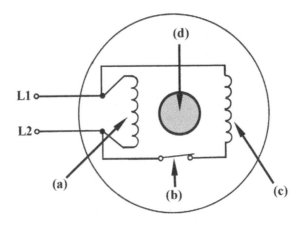

Figure 5-17   Split-phase motor for Question 4.

5. The current flow through the start and run windings of a split-phase motor is in phase with each other. (True/False) _____

6. The direction of rotation on a split-phase motor can be reversed by interchanging the connections to

    a. The run winding

    b. The start winding

    c. Either a or b

    d. Both a and b

    Answer _____

7. Connect the dual-voltage split-phase motor shown in Figure 5-18 for 230-V CCW operation.

Figure 5-18 Dual-voltage, split-phase motor for Question 7.

8. Compared to split-phase motors, capacitor-start motors provide increase starting _____ with less starting _____.

9. Complete the internal wiring of the capacitor-start motor shown in Figure 5-19 to the supply lines.

Figure 5-19 Capacitor-start motor for Question 9.

10. With capacitor-type motors, the capacitor can be a source of trouble if it becomes _____ or _____-circuited.

11. A permanent-capacitor motor

    a. Does not require a centrifugal switch

    b. Uses identical run and auxiliary windings

    c. Provides improvement of motor power factor

    d. All of these

    Answer _____

12. Complete the wiring for the reversing permanent-capacitor motor circuit in Figure 5-20.

Figure 5-20   Reversing motor circuit for Question 12.

13. Capacitor-start/capacitor-run motors use two identical capacitors. (True/False) _____

14. The capacitor-start/capacitor-run motor has the lowest starting torque of all the capacitor type motors. (True/False) _____

15. Complete the internal wiring of the capacitor-start/capacitor run motor shown in Figure 5-21 to the supply lines.

Start Capacitor

Run Capacitor

Start Switch

L1 o

Run Winding

L2 o

Auxiliary Winding

Figure 5-21   Motor for Question 15.

16. Identify the major components of the shaded-pole motor shown in Figure 5-22.

    a. _____

    b. _____

    c. _____

Figure 5-22

Shaded-pole motor

for Question 16.

17. Shaded-pole motors are used to drive

    a. Compressors

    b. Woodworking machinery

    c. Small fans

    d. All of these

    Answer _____

18. Identify the major components of the universal motor shown in Figure 5-23.

    a. _____

    b. _____

    c. _____

Figure 5-23

Universal motor

for Question 18.

19. A universal motor is designed to operate from

    a. An AC single-phase supply

    b. A three-phase supply

    c. A DC supply

    d. Both a and c

    Answer _____

20. Complete the internal wiring of the universal motor shown in Figure 5-24 to the supply lines.

Figure 5-24

Universal motor for Question 20.

L1 O
L2 O

21. Reversing the direction of rotation of a universal motor is accomplished by interchanging the two line leads. (True/False) _____

PART 5   Alternating Current Motor Drives

1. A variable-frequency drive is a system for controlling the rotational speed of an AC motor by controlling the _____ of the electrical power supplied to the motor.

2. The higher the frequency of the power supplied to an AC motor, the _____ (faster/slower) the motor will turn.

3. Identify the major sections of the variable-frequency drive shown in Figure 5-25.

    a. _____

    b. _____

    c. _____

Figure 5-25 Variable-frequency drive for Question 3.

4. The inverter section of a variable-frequency drive rectifies the incoming AC power and converts it to DC. (True/False) _____

5. If an AC motor is designed to operate at 460 V at 60 Hz, the applied voltage must be reduced to _____ V when the frequency is reduced to 15 Hz.

6. The motor voltage waveform shown in Figure 5-26 is an example of _____ voltage control.

    a. Pulse-width modulated

    b. Direct current

    c. Sine wave

    d. Indirect

    Answer _____

Figure 5-26 Voltage waveform for Question 6.

7. _____ duty describes a class of motors specifically designed for use with variable-frequency drives.

    a. Heavy

    b. Light

    c. Inverter

    d. Constant

    Answer _____

PART 6   Motor Selection

1. Slower motors are generally smaller, lighter, and less expensive than faster motors of equivalent horsepower rating. (True/False) _____

2. Motors that are not fully loaded draw _____ (more/less) current than rated nameplate current.

3. Calculate the maximum locked-rotor current for the following three-phase squirrel-cage motor: NEMA Code letter C (4.0 kVA/hp), 460 V, 100 hp. Answer _____

4. The NEMA _____ letter denotes the motor's performance characteristics relating to torque, starting current, and slip.

5. Motor efficiency is the ratio of _____ power output to the _____ power input, usually expressed as a percentage.

6. The core, copper, and mechanical losses associated with the operation of a motor are dissipated as _____ through the body of the motor.

7. Frame sizes are used to ensure the interchangeability of motors having the same frame size. (True/False) _____

8. If a motor with a nameplate rating of 50 Hz is operated from a 60-Hz power supply, the speed of the motor would
    a. Increase slightly
    b. Decrease slightly
    c. Increase by about 20 percent
    d. Decrease by about 20 percent
    Answer _____

9. A constant horsepower load requires high torque at low speeds and low torque at high speeds. (True/False) _____

10. A _____ is an example of a constant torque load.
    a. Conveyor
    b. Fan
    c. Pump
    d. Lathe
    Answer _____

11. The standard motor ambient temperature rating is

    a. 64°F

    b. 75°F

    c. 85°F

    d. 104°F

    Answer _____

12. What is the safe maximum operating temperature for a motor rated for an ambient temperature of 40°C, temperature rise of 105°C, and a hot spot allowance of 10°C?

    Answer _____

13. Continuous-duty-cycle motors are labeled INTER on the nameplate. (True/False) _____

14. The amount of torque produced by a motor when it is initially energized at full voltage is known as the

    a. Full-load torque

    b. Locked-rotor torque

    c. Breakdown torque

    d. Pull-up torque

    Answer _____

15. An _____ motor enclosure has ventilating openings that permit passage of external air over and around the motor windings.

16. Hazardous location motors are designed for environments where explosive or ignitable _____ or _____ are present or may become present.

17. Metric motors are rated in _____ rather than horsepower.

18. Which of the following symbols represents a European motor standard?

    a. UL

    b. CSA

    c. CE

    d. UCAL

    Answer _____

PART 7   Motor Installation

1. Which of the following types of motor mountings will allow direct mounting with a pump?
    a. Adjustable base
    b. Rigid base
    c. Resilient base
    d. NEMA C face mount
    Answer _____

2. Moving a motor or placing a shim under the feet of the motor is not normally part of the alignment process. (True/False) _____

3. Determine the speed of the load pulley for the belt drive system shown in Figure 5-27.
    Answer _____

Motor Drive Pulley
3450-rpm
4-inch diameter

Load Pulley
12-inch diameter

Figure 5-27   Belt drive system for Question 3.

4. Figure 5-28 illustrates an example of _____ misalignment.

Figure 5-28   Belt drive
for Question 4.

5. Motors come equipped with _____ properly lubricated to prevent metal-to-metal contact of the motor shaft.

6. Article _____ of the National Electrical code (NEC) contains recommendations covering motors and motor controls.

7. The equipment grounding conductor for motor circuits may be

   a. An insulated conductor run with the circuit conductors

   b. A bare conductor run with the circuit conductors

   c. The metal raceway containing the circuit conductors

   d. All of these

   Answer _____

8. The color _____ is reserved for an insulated grounding conductor.

9. Bearing current is the result of induced voltage in the motor rotor seeking a path to ground through the motor bearings. (True/False) _____

10. Given: a single 10-hp, three-phase, 208-V squirrel-cage motor. According to the NEC, determine each of the following:

    a. Full-load current (FLC) of the motor _____

    b. Minimum ampacity of the branch circuit conductors _____

    c. Branch circuit wire size for type THWN CU conductors _____

11. For the three-phase voltage supply illustrated in Figure 5-29, determine each of the following:

    a. Average voltage _____

    b. Maximum deviation from the average voltage _____

    c. Percent voltage unbalance _____

Figure 5-29 Three-phase supply for Question 11.

12. Unbalanced motor voltages applied to a polyphase induction motor may cause unbalanced currents resulting in motor overheating. (True/False)

_____

13. What type of built-in thermal motor protection would most likely be used for roof ventilation fan motor application?
    a. Normally open type
    b. Totally enclosed type
    c. Automatic reset type
    d. Manual reset type
    Answer _____

PART 8   Motor Maintenance and Troubleshooting

1. Two things unique to DC motor maintenance are _____ and _____ care.

2. The typical minimum motor insulation resistance for a motor rated for 600 V or less is
    a. 50 Ω
    b. 500 Ω
    c. 1.5 kΩ
    d. 1.5 MΩ
    Answer _____

3. Motors that are used continuously are prone to moisture problems. (True/False) _____

4. Excessive starting of a motor should be avoided because high current flow during start-up generates a great amount of _____ within the motor.

5. Identify the major parts of the motor system shown in Figure 5-30.

   a. _____

   b. _____

   c. _____

   d. _____

Figure 5-30

Motor system for Question 5.

6. Before beginning electrical troubleshooting of a motor, check for all possible _____ malfunctions.

7. List six common causes for failure of any type of motor to start.

   a. _____

   b. _____

   c. _____

   d. _____

   e. _____

   f. _____

8. A bad motor bearing can cause excessive _____ and _____.

9. A broken or disconnected equipment _____ conductor may cause the motor to produce an electric shock when touched.

10. A split-phase motor hums, and will run normally if started by hand. The most probable cause is

    a. A defective motor thermal overload protector

    b. The centrifugal switch is not operating properly

    c. The run winding is open

    d. The load is too high

    Answer _____

11. The start capacitor on a capacitor-start motor continuously fails. The most probable cause is

    a. The motor is being cycled on and off too frequently

    b. An open circuit in the start winding

    c. The centrifugal switch is faulted open

    d. The motor bearings are worn

    Answer _____

12. A single-phase failure of a three-phase motor is the result of an _____ in one phase of the power supply to the motor.

13. The presence of _____ distortion in the applied voltage to a three-phase motor will increase the motor temperature.

14. When a wound-rotor induction motor fails to start, considerable focus should be given to failed components in the rotor _____ bank.

15. List six common causes of excessive arcing at the brushes of direct current motors.

    a. _____

    b. _____

    c. _____

    d. _____

    e. _____

    f. _____

16. Answer each of the following using the motor nameplate information shown in Figure 5-31.

    a. The motor is a _____ - phase type.

    b. The frame number _____ gives all the critical measurements of the motor.

    c. The normal operating current of the motor when running at rated horsepower and 460 V is _____ A.

    d. The multiplier applied to the motor horsepower is _____.

    e. The maximum operating temperature of the motor windings is indicated by the letter _____.

    f. The rated mechanical output of the motor is _____.

    g. The value of the rated low voltage of the motor is _____ V.

h. The maximum temperature of the surrounding environment that will allow rated horsepower without damage is _____.

i. The letter that refers to the locked-rotor characteristics of the motor is _____.

j. The rated operating frequency of the motor is _____.

k. The input electrical watts converted to mechanical output watts or horsepower is _____.

l. The letter that refers to the motor's operating characteristics such as the starting torque and starting current is _____.

m. The rated speed of the motor when operated at 230 V and full load is _____.

n. The length of time the motor can run at full load without overheating is _____.

Figure 5-31 Nameplate for Question 16.

17. Which test instrument would be most likely used to measure the capacitance of a capacitor?

    a. Digital multimeter

    b. Megohmmeter

    c. Tachometer

    d. Oscilloscope

    Answer _____

18. Which test instrument would be most likely used to measure the speed of motor?

    a. Digital multimeter

    b. Megohmmeter

    c. Tachometer

    d. Oscilloscope

    Answer _____

19. Which test instrument would be most likely used to measure the insulation resistance of a motor winding?

    a. Digital multimeter

    b. Megohmmeter

    c. Tachometer

    d. Oscilloscope

    Answer _____

20. Which test instrument would be most likely used to measure the current of a motor?

    a. Clamp-on ammeter

    b. Digital multimeter

    c. Infrared thermometer

    d. Oscilloscope

    Answer _____

21. Which test instrument would be most likely used to pinpoint a hot point on the surface of a motor?

    a. Clamp-on ammeter

    b. Digital multimeter

    c. Infrared thermometer

    d. Oscilloscope

    Answer _____

22. Which test instrument would be most likely used to examine the output voltage waveform from a motor drive?

a. Clamp-on ammeter

b. Digital multimeter

c. Infrared thermometer

d. Oscilloscope

Answer _____

"THE CONSTRUCTOR" CIRCUIT ANALYSIS ASSIGNMENTS

1. The purpose of this assignment is to observe the operation of a DC reversing motor starter used to operate a shunt motor in the forward and reverse directions. Download circuit file Chapter 5 Simulation Assignment 01 (Figure 5-32, p. 100) from the assignment diagrams folder. Click on the Main Power Switch (ON/OFF) located on the left side of the toolbar. Left-click on the circuit breaker to close it. Left-click on the directional push buttons to operate the motor in the forward and reverse directions.

a. How is reversing of the motor accomplished?

_____

b. Assume the motor is running in the forward direction and the reverse push button is closed. What happens as a result? Why?

_____

_____

c. Close the forward and reverse power contact by left-clicking on them. This will simulate a fault condition resulting from both the forward and reverse contactors being energized at the same time. Click the main circuit breaker on. What would happen? Why?

_____

_____

Figure 5-32 DC reversing motor starter.

2. The purpose of this assignment is to observe the operation of an AC reversing motor starter used to operate a three-phase squirrel-cage motor in the forward and reverse directions. Download circuit file Chapter 5 Simulation Assignment 02 (Figure 5-33, p. 101) from the assignment diagrams folder. Click on the Main Power Switch (ON/OFF) located on the left side of the toolbar. Left-click on the circuit breaker to close it. Left-click on the directional push buttons to operate the motor in the forward and reverse directions.

a. How is reversing of the motor accomplished?

_____

b. When the motor is operating in the forward direction, L1, L2, and L3 connect to motor leads _____ respectively.

c. When the motor is operating in the reverse direction, L1, L2, and L3 connect to motor leads _____ respectively.

Figure 5-33 AC reversing motor starter.

6 Contactors and Motor Starters

PART 1   Magnetic Contactor

Place the answers in the space provided.

1. The magnetic contactor is similar in operation to the electromechanical

    _____.

2. Contactors are designed to make and break electric circuit loads of higher

    currents than relays. (True/False) _____

3. Identify the components of the contactor shown in

    Figure 6-1.

    a. _____

    b. _____

    c. _____

Figure 6-1   Contactor for Question 3.

4. The coil is part a magnetic contactor's _____ circuit.

5. The contacts of a contactor are activated any time voltage is applied to the

    a. Coil

    b. Power contacts

    c. Auxiliary contacts

    d. All of these

    Answer _____

6. A _____-pole contactor would be required to switch power to a

    three-phase load.

7. Identify the line side, load side, and parts of the contactor shown in Figure 6-2.

a. _____

b. _____

c. _____

d. _____

e. _____

Figure 6-2   Contactor for Question 7.

8. Complete the wiring for the circuit drawn in Figure 6-3 connected so that the ON/OFF pilot switch switches the current to the three-phase heater load.

Figure 6-3   Circuit for Question 8.

9. Contactors may be used for switching motor loads when separate _____ protection is provided.

10. If the control circuit is connected directly to the same line wires as the load, then it must be rated for a lower voltage. (True/False) _____

11. Pilot lights operating in conjunction with a contactor are normally controlled by auxiliary contacts. (True/False) _____

12. With an _____ held contactor, the coil needs to be energized continuously all the time the main contacts are closed.

13. A mechanically held contactor

    a. Requires only a pulse of coil current to change state.

    b. Operates more efficiently than electrically held types.

    c. Is quieter than electrically held types.

    d. All of these

    Answer _____

14. Complete the wiring for the lighting contactor circuit drawn in Figure 6-4 with the additional ON/OFF control station wired into the circuit.

Figure 6-4   Lighting contactor circuit for Question 14.

15. The type of contactor operating mechanism shown in Figure 6-5 is the

    a. Clapper type

    b. Bell-crank type

    c. Seal-in type

    d. Horizontal-action type

    Answer _____

Figure 6-5   Contactor mechanism for Question 15.

16. A defective contactor coil will read zero or infinity, indicating a

_____ or _____ coil, respectively.

17. Contactor coils are designed to operate

    a. On AC only

    b. On DC only

    c. Over a range of 85 to 110 percent of rated voltage

    d. At rated voltage only

    Answer _____

18. Operating contactor coils at either lower or higher than their rated voltage may result in a high level of contact bounce. (True/False) _____

19. The hold-in voltage rating of a contactor coil is normally

    a. The same as the pickup voltage rating

    b. Less than the pickup voltage rating

    c. Greater than the pickup voltage rating

    d. The same as the dropout voltage rating

    Answer _____

20. AC and DC contactor coils with the same voltage ratings are normally interchangeable. (True/False) _____

21. Current flow through an AC coil is limited by both _____ and

_____.

22. An AC contactor coil has a

    a. High ohmic resistance

    b. High inrush current

    c. A nonlaminated steel core and armature

    d. All of these

    Answer _____

23. Complete the wiring for the PLC contactor-operated circuit drawn in Figure 6-6 with an *RC* suppression module wired into the circuit.

Figure 6-6 Circuit for Question 23.

24. Whenever current to a contactor coil is turned off, a high-voltage spike is generated. (True/False) _____

25. Identify the parts of the contactor assembly shown in Figure 6-7.

a. _____

b. _____

Figure 6-7 Contactor assembly for

Question 25.

26. The core and armature of AC contactor assemblies are made of laminated steel to

a. Reduce chatter

b. Reduce the size of the contactor

c. Reduce eddy-current flow

d. All of these

Answer _____

106

27. A broken or open shading coil will cause a contactor to become extremely noisy. (True/False) _____

28. A contactor fails to seat properly when energized. This may result in

    a. Hum coming from the contactor

    b. Higher than normal current flow through the coil

    c. Overheating of the contactor coil

    d. All of these

    Answer _____

29. Contacts are silver-coated to reduce contact _____.

30. Auxiliary contacts are used to switch high-current load circuits. (True/False) _____

31. Silver contacts should never be filed. (True/False) _____

32. Arc current flow produces additional heat, which can damage _____ surfaces.

33. On opening under load, as the distance between contacts increases, the

    a. Resistance of the arc remains the same

    b. Resistance of the arc increases

    c. Voltage required to sustain remains the same

    d. Voltage required to sustain the arc deceases

    Answer _____

34. An AC arc is self-extinguishing, as the arc will normally extinguish as the AC cycle passes through _____.

35. A motor would be classified as a

    a. Resistive load

    b. Nonlinear load

    c. Inductive load

    d. Capacitive load

    Answer _____

36. Identify the parts of the contactor shown in Figure 6-8.

    a. _____

    b. _____

Figure 6-8 Contactor for Question 36.

37. Blowouts are connected in series with the contacts. (True/False) _____

38. Worn contacts should always be replaced in pairs. (True/False) _____

39. The contactor shown in Figure 6-9 would be

    classified as a _____ type.

Figure 6-9 Contactor for Question 39.

40. A vacuum provides a better environment than free air for breaking the arc

    produced by opening contacts under load. (True/False) _____

41. A vacuum contactor does not use a coil to operate the contacts.

    (True/False) _____

42. Contactor auxiliary contacts have a current rating much higher than that of the

    main contacts. (True/False) _____

PART 2   Contactor Ratings, Enclosures, and Solid-State Types

1. The NEMA size categorizes a contactor on the basis of _____,

    _____, and _____.

2. Which of the following NEMA size contactors would have the greatest current-

    carrying capacity?

    a. 0

    b. 3

    c. 6

    d. 9

    Answer _____

3. Name the contactor IEC load utilization category for each of the following:

    a. Transformer _____

    b. Incandescent lighting installation _____

    c. Electric furnace heater _____

    d. Power factor correction bank _____

4. NEMA contactors, as compared to IEC types, generally are physically downsized

    to provide higher ratings in a smaller package. (True/False) _____

5. NEMA contactors are categorized by _____, while IEC types are categorized by _____ categories.

6. The two general classifications for NEMA contactor enclosures are _____ location and _____ location.

7. Which of the following NEMA-type enclosures would be selected to ensure that an internal explosion would be confined to the inside of the enclosure?

    a. Type 1

    b. Type 4X

    c. Type 7

    d. Type 12

    Answer _____

8. The IEC method for specifying the enclosures of electrical equipment is the same as that used by NEMA. (True/False) _____

9. Solid-state switching refers to interruption of power by _____ means.

10. A solid-state contactor

    a. Has no contacts

    b. Is silent in operation

    c. Produces no arc

    d. All of these

    Answer _____

11. Identify the leads of the SCR solid-state switch shown in Figure 6-10.

    a. _____

    b. _____

    c. _____

    (b) ——▶|—— (c)
    (a)

Figure 6-10 Solid-state switch for Question 11.

12. With reference to the SCR testing circuit shown in Figure 6-11, identify the SCR fault, if any, commonly associated with each of the following scenarios:

    a. Light turns on as soon as the power is applied.

    _____

    b. Light turns on when the ON button is pressed and remains on when the button is released.

    _____

c. Light fails to turn on when the ON button is pressed and power is applied.

Figure 6-11 Testing circuit for Question 12.

13. Zero-fired control of an SCR refers to triggering the SCR on at the maximum point of the sine wave so that no current is being switched under load. (True/False) _____

14. Any device that makes use of a coil of wire for its operation can be classed as a(n) _____ load.
    a. Capacitive
    b. Resistive
    c. Passive
    d. Inductive
    Answer _____

15. Which of the following loads can create voltage spikes?
    a. Transformers
    b. Relay coils
    c. Solenoids
    d. All of these
    Answer _____

16. Unlike magnetic contactors, solid-state types can be switched ON and OFF by a lower-power digital control signal. (True/False) _____

PART 3   Motor Starters

1. Complete the magnetic motor starter wiring diagram in Figure 6-12 according to the control ladder diagram. Do not make any wire splices or additional terminal connections on the wiring diagram. All connections must run from terminal screw to terminal screw.

Figure 6-12 Magnetic starter for Question 1.

2. In its most common form, a motor starter is a contactor with

    a. An overload protection device

    b. An armature

    c. A circuit breaker

    d. A push button

    Answer _____

3. Enclosures protect people from electric shock. (True/False) _____

4. Identify the major NEC requirements for the motor installation shown in Figure 6-13.

   a. _____

   b. _____

   c. _____

   d. _____

   e. _____

5. The starting current of a motor is normally several times higher than the rated full-load nameplate current. (True/False) _____

6. A Class _____ overload relay will trip an overloaded motor offline within 20 seconds at six times full-load amperes.

   a. 20

   b. 10

   c. 40

   d. 35

   Answer _____

Figure 6-13 Motor installation for Question 4.

7. Once tripped, overload relays normally require a cooling-off period before they can be reset. (True/False) _____

8. Overload relays can be classified as being _____, _____, or _____.

9. External overload protection devices, which are mounted in the starter, attempt to simulate the heating and cooling of a motor by sensing the _____ flowing to it.

10. A thermal overload relay uses a heater element connected in parallel with the motor supply. (True/False) _____

11. A thermal overload relay protecting a three-phase motor requires _____ heaters.

   a. 1

   b. 2

   c. 3

   d. 4

   Answer _____

12. Identify the parts of the thermal-overload relay shown in Figure 6-14.

   a. _____

   b. _____

Figure 6-14 Relay for Question 12.

13. Identify the parts of the eutectic overload relay shown in Figure 6-15.

   a. _____

   b. _____

Figure 6-15 Eutectic overload relay for Question 13.

14. Identify the parts of the bimetallic overload relay shown in Figure 6-16.

   a. _____

   b. _____

   c. _____

Figure 6-16 Bimetallic overload relay for Question 14.

15. Selection tables for thermal overload relays normally list OL heaters according to motor _____.

16. Ambient temperatures will not normally affect the tripping time of a thermal overload relay. (True/False) _____

17. Normally, in selecting OL heater sizes from a manufacturer's table, it is assumed that the motor
    a. Has a minimum service factor of 1.1
    b. Temperature rise is not over 104°F
    c. Will be protected up to 125 percent of the nameplate FLC rating
    d. All of these
    Answer _____

18. An electronic overload relay measures motor current directly through a current _____.

19. Electronic overload relays
    a. Can be connected directly to the contactors
    b. Are easily adjustable to a wide range of full-load motor currents
    c. Reduce the amount of heat generated by the starter
    d. All of these
    Answer _____

20. For the electronic overload relay shown in Figure 6-17, identify the function or connection of each part.

    a. _____
    _____

    b. _____
    _____

    c. _____
    _____

    d. _____
    _____

    e. _____

    f. _____

Figure 6-17 Electronic overload relay for Question 20.

21. Dual-element fuses, when properly sized, provide

    a. Short-circuit protection

    b. Ground fault protection

    c. Overload protection

    d. All of these

    Answer _____

22. Identify the elements of the dual-element fuse shown in Figure 6-18.

    a. _____ (a)

    b. _____

Figure 6-18   Dual-element fuse for

Question 22.   (b)

23. Identify each type IEC motor control symbol shown in Figure 6-19.

a. _____   b. _____   c. _____

d. _____   e. _____   f. _____

Figure 6-19   IEC symbols for Question 23.

"THE CONSTRUCTOR" CIRCUIT ANALYSIS ASSIGNMENTS

1. The purpose of this assignment is to observe the operation of a magnetic contactor control of a heater load. Download circuit file Chapter 6 Simulation Assignment 01 (Figure 6-20) from the assignment diagrams folder. Click on the Main Power Switch (ON/OFF) located on the left side of the toolbar. Left-click on the circuit breaker to close it. Left-click on the heater operating control devices to turn the heater on and off.

   a. The value of the power circuit voltage is _____ V.

   b. The value of the control circuit voltage is _____ V.

   c. The value of the contactor coil voltage is _____ V.

Figure 6-20 Magnetic contactor control of a heater load.

d. When on, what do the green and red pilot lights indicate?

_____

_____

e. Left-click one of the heaters in the three-phase bank to open-circuit it. In what way does this affect the control and power circuit?

_____

_____

f. Right-click on any two of the heaters in the three-phase bank to short-circuit them. In what way does this affect the control and power circuit?

_____

_____

g. What type of contacts would CR-4, CR-5, and CR-6 be classified as?

_____

h. A high-temperature sensor, equipped with a single normally closed contact, is to be wired into the circuit to prevent the heaters from being energized should the temperature rise above a preset safe value. Explain how this sensor would be connected into the circuit. Switch to the design mode and connect the temperature sensor into the circuit. Simulate the operation of the circuit with the sensor connected.

_____

_____

2. The purpose of this assignment is to observe the operation of a dual-coil mechanically held lighting contactor. Download circuit file Chapter 6 Simulation Assignment 02 (Figure 6-21, p. 118) from the assignment diagrams folder. Click on the Main Power Switch (ON/OFF) located on the left side of the toolbar. Left-click on the main circuit breaker. Turn the branch circuit light banks on and off.

a. The two lighting circuits are _____-phase and generally rated at _____ V or _____ V.

b. With the main circuit breaker on, what procedure is followed to turn both banks of lights on?

_____

c. Whenever both banks of lights are on, what is the state (open/closed) of contacts M3 and M4?

_____

d. Whenever both banks of lights are on, what is the value of the current flow through the contactor coil?

_____

e. To what points does the common lead of the latch/unlatch coils connect?

_____

Figure 6-21 Dual-coil mechanically held lighting contactor.

7 Relays

PART 1   Electromechanical Control Relays

Place the answers in the space provided.

1. The two basic parts of an electromechanical relay are the _____ and the _____ .

2. The two circuits associated with a relay are the

   a. Primary and secondary

   b. Control and load

   c. Normally open and normally closed

   d. High and low

   Answer _____

3. Complete the relay wiring diagram for Figure 7-1 to operate so that opening and closing the switch will control current to the load through the relay contact.

Figure 7-1

Wiring
diagram for
Question 3.

119

4. Relays are used to control large loads of 15 A or more. (True/False) _____

5. An electromechanical relay normally will have only one _____, but it may have a number of different _____.

6. Normally closed contacts are defined as those contacts that are closed when the coil is _____

7. State the correct type (N.O. or NC) and label for each of the contacts associated with the relay coil of Figure 7-2.

   a. _____

   b. _____

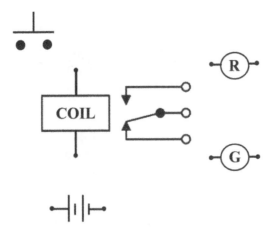

Figure 7-2   Relay coil for Question 7.   (a)   (b)

8. Which of the following is *not* a common application for a relay?

   a. To control a 120 V lighting circuit with a 12 V control circuit.

   b. To control several switching operations by a single separate current.

   c. To change alternating current to direct current.

   d. To switch a small servo motor with a low-current sensor signal.

   Answer _____

9. The coil and contacts of a relay are not normally electrically insulated from each other. (True/False) _____

10. Complete the relay wiring diagram for Figure 7-3 to operate so that when the push button is closed, the green light switches from on to off and the red light switches from off to on.

Figure 7-3   Relay wiring diagram for Question 10.

11. Relay options that aid in troubleshooting may include:

    a. _____

    b. _____

12. It is possible for a relay coil to be rated for 24 V DC and its contacts to be rated for 240 V AC. (True/False) _____

13. Electromechanical relay contacts may be constructed as

    a. N.O. or NC contacts.

    b. Single-break or double-break contacts.

    c. Fixed and movable contacts.

    d. All of these

    Answer _____

14. State the number of poles, throws, and breaks for each of the relay contacts shown in Figure 7-4.

    a. _____-pole

    _____-throw

    _____-break

    b. _____-pole

    _____-throw

    _____-break

    c. _____-pole

    _____-throw

    _____-break

    d. _____-pole

    _____-throw

    _____-break

Figure 7-4   Relay contacts for Question 14.

15. The load-carrying capacity of contacts is normally given as a current value for a _____.

    a. Resistive load

    b. Inductive load

    c. Capacitive load

    d. Shorted load

    Answer _____

16. A relay contact rated 12 A at 12 V DC can switch _____ A at 48 V DC.

PART 2   Solid-State Relays

1. In the symbol of the solid-state relay
    (SSR) shown in Figure 7-5, the #1 set of
    leads is comparable to the
    _____ leads of an
    electromechanical relay and the #2 set to
    the _____ leads.

Figure 7-5 Symbol for Question 1.

2. List three common main switching semiconductors used in solid-state relays.

    a. _____

    b. _____

    c. _____

3. A common method used to provide isolation is to have the input section

    illuminate a _____ that activates a photodetective device.

    a. LED
    b. SCR
    c. Transistor
    d. Triac

    Answer _____

4. Identify the components of the SSR shown in Figure 7-6.

    a. _____

    b. _____

    c. _____

    d. _____

Figure 7-6   Solid-state relay for Question 4.

5. The input section of a SSR acts like the contacts of an electromagnetic relay (EMR). (True/False) _____

6. The use of solid-state relays is not recommended for loads that must be switched continually and quickly. (True/False) _____

7. Most SSRs have a variable input voltage range. (True/False) _____

8. A majority of SSRs are multiple-pole devices. (True/False) _____

9. Complete the solid-state relay wiring diagram in Figure 7-7 to operate so that when the control contact is closed, three-phase power is delivered to the load.

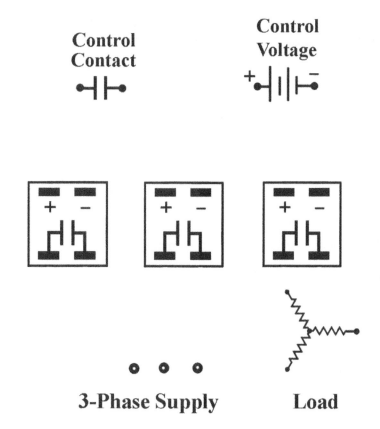

Figure 7-7 Wiring diagram for Question 9.

10. Complete the three-wire motor control diagram for Figure 7-8 utilizing the SCR and solid-state relay. The circuit is to operate so that the starter coil is energized when the start button is momentarily pressed and deenergized whenever either of the two stop buttons is momentarily pressed.

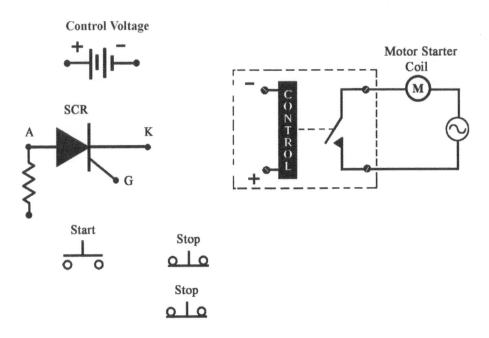

Figure 7-8   Motor control diagram for Question 10.

11. What type of SSR switching is primarily designed to work with resistive loads?
    a. Instant-on switching
    b. Peak switching
    c. Phase-shift switching
    d. Zero switching
    Answer _____

12. _____ SSR switching initiates the output at the exact same time the input signal is received.

13. Zero switching of resistive loads increases the inrush current.
    (True/False) _____

14. Compared to electromechanical relays, solid-state relays

    a. Do not exhibit contact bounce

    b. Have a much faster response time

    c. Do not generate as much electromagnetic interference

    d. All of these

    Answer _____

15. Compared to EMR contacts, the SSR switching semiconductor has a much lower on-state resistance. (True/False) _____

PART 3   Timing Relays

1. Timing relays are used to _____ the opening or closing of contacts for circuit control.

2. Contacts of a synchronous clock timer are tripped by

    a. Hand

    b. An electromagnet

    c. Tabs set along a timing wheel

    d. A control signal

    Answer _____

3. Synchronous clock timers are not accurate enough for critical timing processes. (True/False) _____

4. Timing is managed in a dashpot pneumatic timer by controlling _____ flow through a small orifice.

5. Pneumatic timers have a relatively narrow timing range. (True/False) _____

6. Compared with dashpot timers, solid-state timers

    a. Provide much more accurate timing

    b. Have a narrower timing range

    c. Are initiated by an electromagnetic coil

    d. All of these

    Answer _____

7. Complete the solid-state timer wiring diagram for Figure 7-9. The circuit is to operate so that load is energized at 120 V AC following the closure of the float switch and preset time delay period.

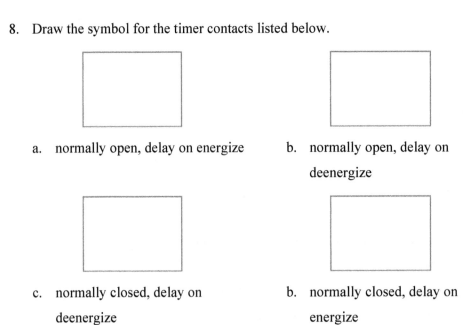

Figure 7-9   Wiring diagram for Question 7.

8. Draw the symbol for the timer contacts listed below.

a.   normally open, delay on energize

b.   normally open, delay on deenergize

c.   normally closed, delay on deenergize

b.   normally closed, delay on energize

9. Examine the timing diagrams of Figure 7-10 and identify the timer type as DOE, DODE, one-shot, or recycle.

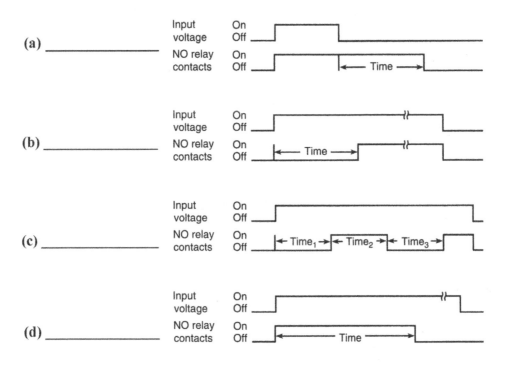

(a) _____

(b) _____

(c) _____

(d) _____

Figure 7-10 Timing diagrams for Question 9.

10. Answer the following with reference to the timer circuit of Figure 7-11.

a. What is the state (open or closed) of contact TR1-1 prior to the switch being closed?

Answer _____

b. What is the state of contact TR1-2 immediately after the switch is closed.

Answer _____

c. What is the state of contact TR1-1 after 5 seconds of the switch being closed?

Answer _____

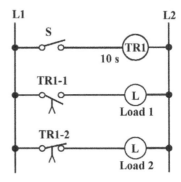

Figure 7-11 Circuit for Question 10.

d. What is the state of contact TR1-2 after 10 seconds of the switch being closed?

Answer _____

e. What is the state of contact TR1-1 as soon as the switch is opened?

Answer _____

11. Answer the following with reference to the timer circuit of Figure 7-12.

   a. What is the state (open or closed) of contact TR1-2 prior to the switch being closed?

   Answer _____

   b. What is the state of contact TR1-1 immediately after the switch is closed?

   Answer _____

Figure 7-12 Circuit for Question 11.

   c. What is the state of contact TR1-2 immediately after the switch is opened?

   Answer _____

   d. What is the state of contact TR1-1 after 10 seconds of the switch being opened?

   Answer _____

   e. What is the state of contact TR1-2 after 10 seconds of the switch being opened?

   Answer _____

12. Answer the following with reference to the plug-in timer base wiring diagram of Figure 7-13.

   a. What are the pin numbers for the normally open contact?

   Answer _____

   b. What are the pin numbers for the normally closed contact?

   Answer _____

Figure 7-13 Circuit for Question 12.

   c. Identify the pin numbers where the constant voltage source is to be connected.

   Answer _____

   d. Identify the pin numbers where an external trigger switch is to be connected.

   Answer _____

13. The one-shot timer shown in Figure 7-14 is set for a time-delay period of 5 seconds. For how long would the load be energized for each of the following scenarios?

   a. Initiate button is held closed for 1 second.

   Answer _____

   b. Initiate button is held closed for 5 seconds.

   Answer _____

   c. Initiate button is held closed for 10 seconds.

   Answer _____

Figure 7-14 Timer for Question 13.

   d. Initiate button is held closed for 2 seconds but power is lost after 1 second.

   Answer _____

14. Answer the following with reference to the recycle timer shown in Figure 7-15.

   a. What type of recycle timer (symmetrical or asymmetrical) is this?

   Answer _____

   b. How is the on period timing adjustment accomplished?

   Answer _____

   c. How is the off period timing adjustment accomplished?

   Answer _____

   d. How is the on/off recycling of the contacts stopped?

   Answer _____

Figure 7-15 Timer for Question 14.

15. The term _____ timer is reserved for a group of solid-state timers that perform more than one timing function.

16. A programmable logic controller can be programmed to implement conventional timer functions. (True/False) _____

17. Complete the on-delay ladder logic program of Figure 7-16 to operate so that output Q1 is energized any time both input switches I1 and I3 are closed for a sustained period of 10 seconds or more.

Figure 7-16 Ladder logic program for Question 17.

PART 4 Latching Relays

1. Latching relays use a _____ latch or permanent _____ to hold the contacts in their last energized position.

2. In a latching relay that uses two coils, the coils are identified as the _____ coil and the _____ coil.

3. Latching relays can provide circuit continuity during power failures. (True/False) _____

4. In a two-coil latching relay, the latch coil requires a _____ flow of current to set the latch and hold the relay in the latched position.

    a. Direct

    b. Alternating

    c. Continuous

    d. Pulsed

    Answer _____

5. On electrical diagrams, the contacts of a latching relay are normally shown

    a. In latched position

    b. In unlatched position

    c. Energized

    d. Deenergized

    Answer _____

6. Answer the following with reference to the latching relay circuit of Figure 7-17.

    a. How is the pilot light turned on?

    Answer _____

    b. How is the pilot light turned off?

    Answer _____

    c. Assume the unlatch coil is faulted open. How would this affect the operation of the circuit?

    Answer _____

Figure 7-17 Circuit for Question 6.

7. Single-coil magnetic latching relays are designed to be polarity sensitive. (True/False) _____

8. Alternating relays are used to equalize the _____ _____ of two loads.

9. Answer the following with reference to the alternating relay circuit of Figure 7-18.

   a. What configuration of relay contacts is used?

   Answer _____

   b. What initiates the switch in output between Load A (LA) and Load B (LB)?

   Answer _____

Figure 7-18 Relay circuit for Question 9.

   c. Remote pilot lights are to be connected to the unused contacts of the relay to indicate which load is active. To which two pins would the pilot lights for load A (LA) and load B (LB) be connected?

   Answer _____

10. _____ - _____ alternating relays have the ability to alternate the loads during normal work loads and then operate both when the load is high.

PART 5   Relay Control Logic

1. Identify the two states of the digital signal shown in Figure 7-19.

   a. _____ b. _____

Figure 7-19 Digital signal for Question 1.

2. Relay control logic and ladder diagrams are the primary programming language for _____.

3. The two separate and distinct parts of a motor control circuit are the primary and secondary sections. (True/False) _____

4. The input section of a control circuit provides the

   a. Signals

   b. Source

   c. Actions

   d. Load

   Answer _____

5. Answer the following with reference to the control circuit of Figure 7-20.

   a. List all the component parts that are considered to be inputs.

   Answer

   _____

   _____

   _____

   _____

   b. List all the component parts that are considered to be outputs.

   Answer

   _____

   _____

   _____

   _____

   _____

Figure 7-20 Control circuit for Question 5.

   c. Which output would be energized with power applied and the circuit in the state shown?

   Answer _____

   d. Assume that the contactor coil becomes faulted open. In what way would the normal operation of the circuit be affected when the temperature switch closes?

   Answer _____

   _____

133

6. All input devices can be classified as being loads. (True/False) _____

7. Logic functions describe how inputs interact with each other to control the outputs. (True/False) _____

8. Logic is the ability to make decisions when one or more different factors must be taken into consideration. (True/False) _____

9. Identify the control logic function associated with each of the contact configurations shown in Figure 7-21.

   a. _____ b. _____ c. _____

   d. _____ e. _____ f. _____

Figure 7-21

Configurations for Question 9.

10. State the type of logic function described in each of the following control circuit scenarios.

   a. Two or more inputs are connected in series and they all must be closed to energize the load.

   Answer _____

   b. Two or more inputs are connected in parallel and any one of the inputs can close to energize the load.

   Answer _____

11. Draw an OR logic circuit in which a normally open push button and temperature switch control a resistive heating element.

12. Draw a NOT logic circuit in which a solenoid and pilot light are controlled by a single push button. Both loads are to be energized when the button is not pressed and deenergized when the button is pressed.

13. Draw a combination NOR/OR logic circuit that uses a three-wire control to energize a magnetic motor starter. The motor can be started and stopped from three locations. Include overload protection for the motor.

hands
on PRACTICAL ASSIGNMENTS

1. The purpose of this assignment is to wire different types of relays in simple configurations that demonstrate their basic operating principle.

   a. Complete a circuit diagram for an electromechanical relay configured to operate a lamp load. Wire the circuit in a neat and professional manner.

   b. Complete a circuit diagram for a solid-state relay configured to operate a lamp load. Wire the circuit in a neat and professional manner.

   c. Complete a circuit diagram for a timing relay configured to operate a lamp load. Wire the circuit in a neat and professional manner.

   d. Complete a circuit diagram for an off-delay latching relay configured to operate a lamp load. Wire the circuit in a neat and professional manner.

   e. Complete a circuit diagram for an alternating relay configured to operate a lamp load. Wire the circuit in a neat and professional manner.

2. The purpose of this assignment is to wire, program, and observe the operation of a programmable logic controller (PLC) on-delay timer.

   a. Construct a PLC wiring diagram and program using whatever type PLC is available to you and configured to operate a lamp load.

   b. Complete the hard wiring of the circuit.

   c. Enter and run the program.

"THE CONSTRUCTOR" CIRCUIT ANALYSIS ASSIGNMENTS

1. The purpose of this assignment is to observe the multiple switching operation of a relay. Download circuit file Chapter 7 Simulation Assignment 01 (Figure 7-22, p. 137) from the assignment diagrams folder. Click on the Main Power Switch (ON/OFF) located on the left side of the toolbar. Left-click on the circuit breaker to close it and on the push button to operate it.

a. Record the state of each light with the push button open.

R - _____ , G - _____ , Y - _____

b. Record the state of each light with the push button closed.

R - _____ , G - _____ , Y - _____

c. Which light remains on with the push button open or closed? Why?

_____

_____

d. Which light(s) is/are on when the relay coil is energized?

_____

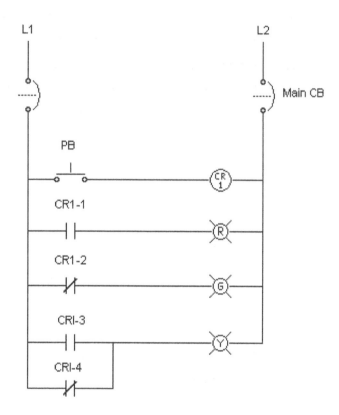

Figure 7-22 Multiple switching relay circuit.

2. The purpose of this assignment is to observe the operation of the instantaneous and delayed switching contacts of a relay. Download circuit file Chapter 7 Simulation Assignment 02 (Figure 7-23) from the assignment diagrams folder. Click on the Main Power Switch (ON/OFF) located on the left side of the toolbar. Left-click on the circuit breaker to close it and on the switch to operate it.

   a. Record the state of each light with the switch open.

      R - _____ , G - _____ , Y - _____ , W- _____

   b. Record the state of each light immediately upon closing the switch.

      R - _____ , G - _____ , Y - _____ , W- _____

   c. Record the state of each light 6 seconds after the switch has been closed.

      R - _____ , G - _____ , Y - _____ , W- _____

   d. Record the state of each light immediately upon opening the switch.

      R - _____ , G - _____ , Y - _____ , W- _____

Figure 7-23   Instantaneous and delayed switching contacts.

3. The purpose of this assignment is to observe the switching operation of an on-delay timer. Download circuit file Chapter 7 Simulation Assignment 03 (Figure 7-24) from the assignment diagrams folder. Click on the Main Power Switch (ON/OFF) located on the left side of the toolbar. Left-click on the circuit breaker to close it and on the switch to operate it.

a. Record the state of each light with the switch open.

R - _____ , G - _____

b. Record the state of each light immediately upon closing the switch.

R - _____ , G - _____

c. Record the state of each light 10 seconds after the switch has been closed.

R - _____ , G - _____

d. Record the state of each light immediately upon opening the switch.

R - _____ , G - _____

Figure 7-24 On-delay timer circuit.

4. The purpose of this assignment is to observe the switching operation of an off-delay timer. Download circuit file Chapter 7 Simulation Assignment 04 (Figure 7-25) from the assignment diagrams folder. Click on the Main Power Switch (ON/OFF) located on the left side of the toolbar. Left-click on the circuit breaker to close it and on the switch to operate it.

   a. Record the state of each light with the switch open.

   R - _____ , G - _____

   b. Record the state of each light immediately upon closing the switch.

   R - _____ , G - _____

   c. Record the state of each light 10 seconds after the switch has been closed.

   R - _____ , G - _____

   d. Record the state of each light immediately upon opening the switch.

   R - _____ , G - _____

   e. Record the state of each light 10 seconds after opening the switch.

   R - _____ , G - _____

Figure 7-25 Off-delay timer circuit.

140

5. The purpose of this assignment is to observe the switching operation of a recycle timer. Download circuit file Chapter 7 Simulation Assignment 05 (Figure 7-26) from the assignment diagrams folder. Click on the Main Power Switch (ON/OFF) located on the left side of the toolbar. Left-click on the circuit breaker to close it and on the switch to operate it.

   a. What type of electrical device is the circuit functioning as?

   _____

   b. What is the value of the on time delay period?

   _____

   c. What is the value of the off time delay period?

   _____

   d. How is the recycling sequence turned on and off?

   _____

   e. What type of symbol is used for the timed contact?

   _____

Figure 7-26   Recycle timer circuit.

6. The purpose of this assignment is to observe the switching operation of a one-shot timer. Download circuit file Chapter 7 Simulation Assignment 06 (Figure 7-27) from the assignment diagrams folder. Click on the Main Power Switch (ON/OFF) located on the left side of the toolbar. Left-click on the circuit breaker to close it and on the push button to operate it.

a. What type of timer (on- or off-delay) is TR1 and TR2?

_____

b. Outline the sequence of operations that occurs when power is first applied and the push button is held continuously closed.

_____

_____

_____

c. Which timer setting determines how long the light remains on?

_____

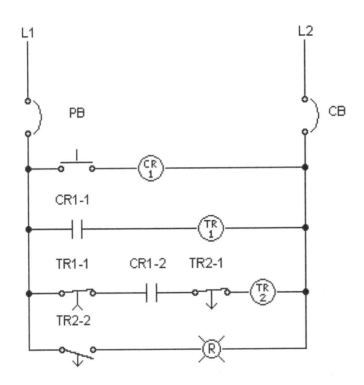

Figure 7-27 One-shot timer circuit.

7. The purpose of this assignment is to observe the operation of a latching relay. Download circuit file Chapter 7 Simulation Assignment 07 (Figure 7-28) from the assignment diagrams folder. Click on the Main Power Switch (ON/OFF) located on the left side of the toolbar. Left-click on the circuit breaker to close it and on the push buttons to operate them. This latching relay is configured using a dual coil with separate set (latch) and reset (unlatch) leads.

a. Momentarily close the latch button and make a note of how the circuit reacts.

Answer _____

b. Momentarily close the unlatch button and make a note of how the circuit reacts.

Answer _____

c. With the pilot light on, open and close the main circuit breaker and make a note of how the circuit reacts.

Answer _____

Figure 7-28 Latching relay circuit.

8. The purpose of this assignment is to observe the operation of an alternating relay. Download circuit file Chapter 7 Simulation Assignment 08 (Figure 7-29) from the assignment diagrams folder. Click on the Main Power Switch (ON/OFF) located on the left side of the toolbar. Left-click on the circuit breaker to close it and on the push button to operate it. Record the status of each pilot light when the following operations are preformed in sequence.

a. Power is first applied.

Answer _____

b. Push button is pressed and released.

Answer _____

c. Push button is pressed and released a second time.

Answer _____

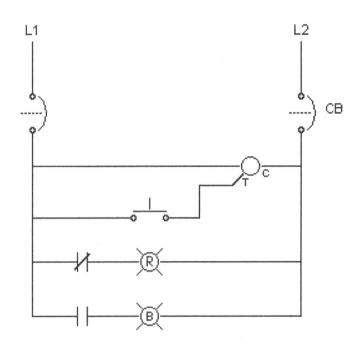

Figure 7-29 Basic alternating relay circuit.

9. The purpose of this assignment is to observe the operation of an alternating relay as part of a motor control circuit. Download circuit file Chapter 7 Simulation Assignment 09 (Figure 7-30) from the assignment diagrams folder. Click on the Main Power Switch (ON/OFF) located on the left side of the toolbar. Left-click on the circuit breaker to close it and on the push buttons to operate them.

   a. State the purpose of the alternating relay used in the circuit

   Answer _____

   _____

   b. Outline the sequence of operations.

   Answer _____

   _____

   _____

Figure 7-30   Alternating relay as part of a motor control circuit.

10. The purpose of this assignment is to identify relay control logic functions. Download circuit file Chapter 7 Simulation Assignment 10 (Figure 7-31) from the assignment diagrams folder. Click on the Main Power Switch (ON/OFF) located on the left side of the toolbar. Left-click on the circuit breaker to close it and on the input control devices to operate the loads. Identify the type of logic function associated with each of the control rungs A to E.

a. Rung A - _____   b. Rung B - _____

c. Rung C - _____   d. Rung D - _____

e. Rung E - _____

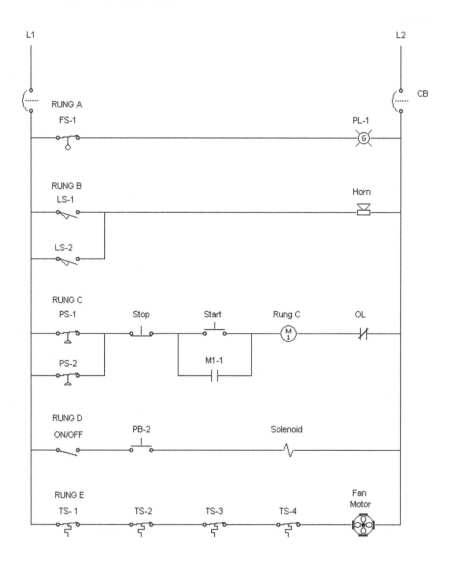

Figure 7-31 Relay control logic functions.

11. The purpose of this assignment is to implement troubleshooting by analysis of the circuit logic. Download circuit file Chapter 7 Simulation Assignment 11 (Figure 7-32, p. 148) from the assignment diagrams folder. Click on the Main Power Switch (ON/OFF) located on the left side of the toolbar. Left-click on the circuit breaker to close it and on the input control devices to operate the loads. Operate the circuit and respond to each of the following troubleshooting scenarios.

   a. Assume the pressure switch is stuck in the open position. How would this affect the operation of the pilot light?

   Answer _____

   _____

   b. Right-click on the solenoid to fault it shorted and operate the limit switches. What happens? Why?

   Answer _____

   _____

   c. Assume proximity switch 2 is fault shorted. How would this affect the operation of the motor starter circuit?

   Answer _____

   _____

   d. Assume the temperature switch is faulted open. How would this affect the operation of the fan motor?

   Answer _____

   _____

   e. Assume that during normal operation and with the fluorescent light bank on, PB3 becomes fault shorted. How would this affect the operation of the light circuit?

   Answer _____

   _____

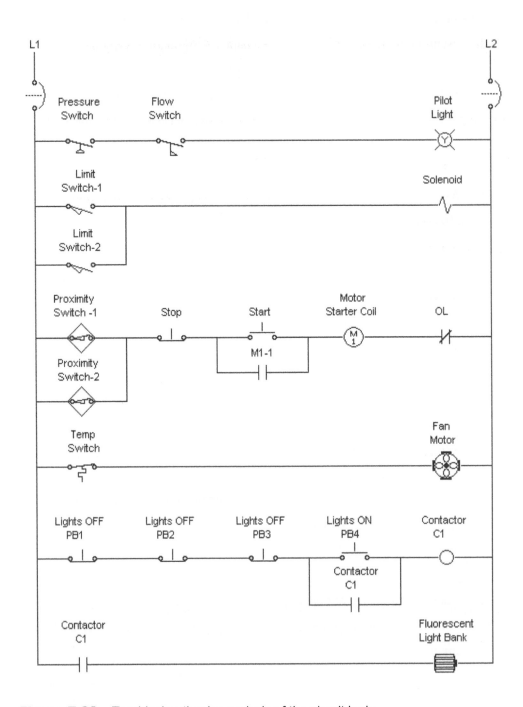

Figure 7-32 Troubleshooting by analysis of the circuit logic.

8 Motor Control Circuits

PART 1   NEC Motor Installation Requirements

Place the answers in the space provided.

1. Article _____ of the National Electrical Code covers installation of

   motor circuits.

2. Using the list of terms provided below, identify by

   letter each requirement for the motor installation shown

   in Figure 8-1.

   a. Overload protection _____

   b. Short-circuit protection _____

   c. Control circuit protection _____

   d. Motor controller _____

   e. Disconnecting means _____

   f. Sizing branch circuit conductors _____

   g. Providing a control circuit _____

Figure 8-1   Motor installation for Question 2.

3. Motor branch circuit conductors must have an ampacity of not less than

   _____ of the motor's full-load current.

   a. 50 percent

   b. 100 percent

   c. 125 percent

   d. 200 percent

   Answer _____

4. The full-load current rating shown on the motor's nameplate is not permitted to be

   used to determine the ampacity of the conductors. (True/False) _____

5. The full-load current rating shown on the motor's nameplate is permitted to be used to determine the overload protection for the motor. (True/False) _____

6. According to the NEC, determine the minimum branch-circuit ampacity required for a 7½-hp, 230-V, three-phase motor.

   Answer _____

7. A 20-hp motor and a 10-hp motor, both 460 V, three phase, share the same feeder. Determine the minimum branch circuit ampacity required to size the feeder conductors.

   Answer _____

8. For the motor branch circuit shown in Figure 8-2, state what undesirable operating condition each of the protective devices guards against.

   a. _____

   b. _____

   c. _____

   d. _____

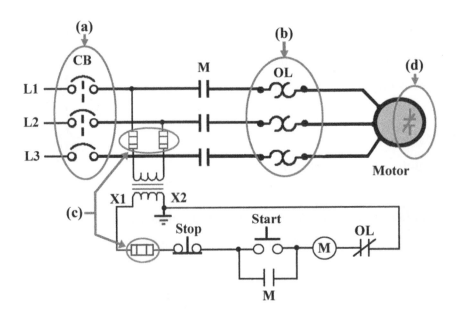

Figure 8-2   Branch circuit for Question 8.

9. The starting current of a motor is approximately the same as its rated full-load current. (True/False) _____

10. With an inverse time circuit breaker, the higher the overcurrent, the longer the time required for the breaker to trip. (True/False) _____

11. According to the NEC, what size of inverse time circuit breaker is permitted to be used for short-circuit and ground fault protection for a 5-hp, 230-V, three-phase, squirrel-cage motor?

    Answer _____

12. A motor overload condition is normally caused by

    a. A ground fault

    b. A short circuit

    c. An open equipment grounding conductor

    d. Excessive motor load

13. Time-delay fuses provide both overload and short-circuit protection. (True/False) _____

14. For any given horsepower motor, the nameplate full-load ampere rating is the same as that given in NEC tables. (True/False) _____

15. In the majority of applications, motor overload protection is provided by overcurrent _____ in the motor controller.

16. Motor control circuit conductors are not permitted to be tapped from the motor branch circuit conductors. (True/False) _____

17. Motor control circuits do not carry the main power current to the motor. (True/False) _____

18. A motor _____ is used to start and stop an electric motor by closing and opening the main power current to the motor.

19. Which of the following is an example of a motor controller?

    a. Magnetic starter

    b. Manual snap switch

    c. Adjustable-speed drive

    d. All of these

    Answer _____

20. The purpose of a motor _____ means is to open the supply conductors to the motor, allowing personnel to work safely on the installation.

21. The disconnecting means for the motor controller and the motor must open all _____ supply conductors simultaneously.

22. In general, each motor requires a disconnecting means that is located within sight of the motor. "In sight of" is defined as being visible and not more than _____ feet in distance from one another.

23. An isolation switch should not be operated under load.

    (True/False) _____

24. Motor control circuits can be the same voltage as the motor up to 600 V or can be reduced by means of a control _____.

25. Where one side of the motor control circuit is grounded, the design of the control circuit must prevent the motor from being started because of a _____ fault in the control circuit wiring.

PART 2   Motor Starting

1. All motors tend to draw much more current during the starting period than when rotating at operating speed. (True/False) _____

2. The current that flows before a motor begins to turn is called its

    a. Locked-rotor current

    b. Full-load current

    c. No-load current

    d. Rated current

    Answer _____

3. An across-the-line starter is designed to apply full line voltage to the motor upon starting. (True/False) _____

4. Manual starters cannot be operated on/off by remotely located control devices.

    (True/False) _____

5. The starter shown in Figure 8-3 would be classified as which of the following types?

    a. Single-pole manual

    b. Double-pole magnetic

    c. Three-pole magnetic

    d. Three-pole manual

    Answer _____

Figure 8-3   Starter for Question 5.

6. On a combination starter, the door cannot be opened while the disconnecting means is closed. (True/False) _____

7. Complete a wiring diagram for the magnetic across-the-line starter shown in Figure 8-4, according to the two start/stop station line diagrams shown in the text. Do not make any wire splices, and run all wire connections from one terminal screw to the other.

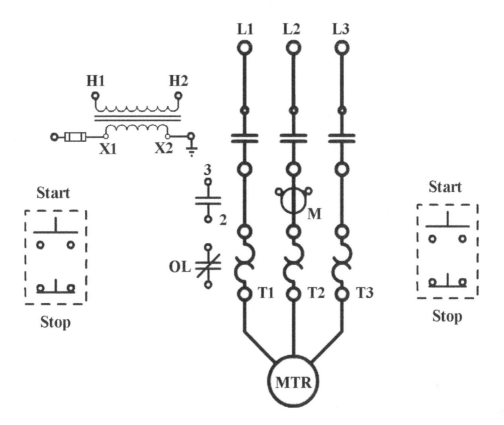

Figure 8-4   Starter for Question 7.

8. A reduced voltage starter
   a. Limits line voltage disturbances
   b. Reduces torque on starting
   c. Increases torque on starting
   d. Both a and b

   Answer _____

9. Answer the following with reference to the primary resistance starter shown in Figure 8-5.

    a. Pressing the start button energizes coils _____ and

        _____.

    b. The motor is initially started through a _____ in each of the incoming lines.

    c. After a preset time delay period, the _____ contact closes to energize coil _____, the contacts of which close to apply full voltage to the motor.

Figure 8-5   Starter for Question 9.

10. Answer the following with reference to the autotransformer starter shown in Figure 8-6.

    a. The instantaneous contacts of timer coil TR are _____ and

        _____, while the timed contacts are _____ and

        _____.

b. The net result of initially pressing the start button is to energize coils

_____, _____, and _____.

c. After the timed contacts change state, coils _____ and

_____ deenergize and coil _____ energizes,

connecting the motor directly to the _____, _____ voltage.

Figure 8-6   Starter for Question 10.

11. Wye-delta starting involves connecting the motor windings in a _____ configuration during the starting period and then in _____ after the motor has begun to accelerate.

12. Wye-delta starters can be used with motors where three or six leads of the stator winding are available. (True/False) _____

13. Open transition wye-delta starters
    a. Apply voltage to the motor during transition
    b. Do not apply voltage to the motor during transition
    c. Do not create voltage surges during transition
    d. Both b and c

    Answer _____

14. Part-winding reduced-voltage starters are used on three-phase motors wound for _____-voltage operation.

15. Part-winding starters are designed to start the motor by energizing both sets of windings on starting and then disconnecting one set when the motor nears operating speed. (True/False) _____

16. Which of the following starters increases the voltage gradually as the motor starts?
    a. Autotransformer starter
    b. Solid-state starter
    c. Wye-delta starter
    d. All of these

    Answer _____

17. Soft-start motor starters limit motor starting current and torque by ramping the voltage applied to the motor during the starting. (True/False) _____

18. Which of the following soft-start acceleration modes of operation regulates the maximum starting current to some preset value?
    a. Dual ramp start
    b. Current limit start
    c. Linear speed acceleration
    d. Full-voltage start

    Answer _____

19. Answer the following with reference to the DC motor starter shown in Figure 8-7.

    a.   This starter would be classified as a reduced-voltage type.

        (True/False) _____

    b.   The use of the three power contacts connected in series is designed to reduce contact _____.

Figure 8-7   Starter for Question 19.

20. Answer the following with reference to the DC motor starter shown in Figure 8-8.

    a.   This starter would be classified as an across-the-line type.

        (True/False) _____

    b.   The voltage applied to the L1, L2, and L3 inputs is three-phase AC.

        (True/False) _____

    c.   The function of the SCRs is to _____ the AC input voltage and provide a _____ DC voltage to the armature.

Figure 8-8   Starter for

Question 20.

PART 3   Motor Reversing and Jogging

1. Interchanging any two line leads to a three-phase induction motor will reverse its direction of rotation. (True/False) _____

2. A reversing starter requires _____ set(s) of overload relays.

3. A reversing starter requires two contactors. (True/False) _____

4. In a three-phase reversing starter, the forward and reverse contactors are mechanically interlocked in order to

   a. Allow the motor to accelerate to full speed

   b. Prevent chattering of the starter coil

   c. Prevent a short circuit from being created across the lines

   d. Both b and c

   Answer _____

5. Pushbutton interlocking of reversing motor starters utilizes

   a. A mechanical link between the forward and reverse push buttons

   b. Momentary and maintained switch contacts

   c. Break-make switch contacts

   d. All of these

   Answer _____

6. With pushbutton interlocking, the normally closed contact on the reverse push button acts like another _____ push button in the forward circuit.

7. Answer the following with reference to the reversing starter control circuit shown in Figure 8-9. This circuit incorporates limit switches to limit travel in the forward and reverse directions.

Figure 8-9

Starter for
Question 7.

a. Pressing the forward push button energizes coil _____.

b. Once operating in the forward direction, the motor is stopped by actuation of the _____ push button or the _____ limit switch.

c. Both limit switches would be classified as being normally _____.

d. With both limit switches closed and the motor operating in the forward direction, pressing the reverse push button will change the direction of rotation. (True/False) _____

8. The reversing starter for a single-phase capacitor start motor changes the direction of rotation by interchanging the _____ winding leads, while those of the run winding remain the same.

9. As a result of the operation of the _____ switch, the single-phase capacitor-start motor must be allowed to slow down before you attempt to reverse the direction of rotation.

10. The reversal of a DC motor is accomplished by

   a. Reversing the two line leads to the motor

   b. Reversing the direction of the armature current only

   c. Reversing the direction of the field current only

   d. Either b or c

   Answer _____

11. The speeding up of a motor when unloaded is known as jogging. (True/False) _____

12. When utilizing jogging control

   a. Care must be taken to prevent dynamic braking

   b. The maintaining contact must not be connected

   c. The thermal-overload element must be removed

   d. All of these

   Answer _____

13. Answer the following with reference to the jog circuit shown in Figure 8-10.

    a. This jog circuit would be classified as a control relay type.
        (True/False) _____

    b. When the switch is placed in the _____ position, the maintaining
        circuit is not broken.

    c. The dual-voltage primary of the control transformer is connected for the
        higher of the two input voltage ratings. (True/False) _____

Figure 8-10 Jog circuit for Question 13.

PART 4   Motor Stopping

1. The most common method of stopping a motor is to remove the
    _____ voltage and allow the motor to coast to a stop.

2. Which of the following braking systems applies to stopping a motor by reversing
    it while it is still running in the forward direction?

    a. Dynamic braking

    b. Regenerative braking

    c. Plugging

    d. Jogging

    Answer _____

3. Which of the following braking systems can hold a motor stationary once it has stopped?

    a. Dynamic braking

    b. Eddy-current brake

    c. Friction brake

    d. Both a and b

    Answer _____

4. A _____ starter is required when a motor is to be brought to a stop by means of plugging.

5. Plugging produces more heat in both the motor and starter than allowing a motor to coast to a stop. (True/False) _____

6. A zero-speed or plugging switch prevents the motor from _____ after it has come to a stop.

7. Answer the following with reference to the antiplugging protection circuit shown in Figure 8-11.

    a. In this application the motor can be reversed but not plugged.
    (True/False) _____

    b. With the motor operating in the forward direction, the F zero-speed switch contacts will be _____ and the R zero-speed contacts will be _____ .

    c. With the motor operating in the reverse direction, the F zero-speed switch contacts will be _____ and the R zero-speed contacts will be _____ .

    d. With the motor at zero speed the F zero-speed switch contacts will be _____ and the R zero-speed contacts will be _____ .

Figure 8-11 Antiplugging circuit for Question 7.

161

8. Dynamic braking is achieved by reconnecting a rotating motor to act as a _____ immediately after it is turned off.

9. DC injection braking is a method of braking in which
    a. DC is applied to the stator windings of an AC motor
    b. AC is applied to the stator windings of an AC motor
    c. DC is applied to the rotor windings of an AC motor
    d. AC is applied to the rotor windings of an AC motor
    Answer _____

10. Electromechanical friction brakes rely on fiction between a _____ and _____ brake arrangement.

11. Electromechanical friction braking is applied by energizing and deenergizing the brake _____.

PART 5   Motor Speed

1. The speed of an induction motor depends on the
    a. Number of poles built into the motor
    b. Frequency of the power supplied to it
    c. Type of braking system employed
    d. Both a and b
    Answer _____

2. A single-speed motor has one rated speed at which it runs when supplied with the nameplate voltage and frequency. (True/False) _____

3. Multispeed motors are available in two basic versions: _____ pole and _____ winding.

4. Answer the following with reference to the multispeed motor circuit shown in Figure 8-12.

    a. The multispeed motor shown would be classified as a _____ _____ type.

    b. This motor requires _____ standard three-pole starter units.

    c. The starters must be mechanically and electrically _____ with each other.

    d. Only one set of overload relays is required to ensure protection on each speed range. (True/False) _____

    e. A six-lead induction motor is required for this application. (True/False) _____

Figure 8-12   Multispeed motor circuit for Question 4.

5. A wound-rotor induction motor has a _____ like the squirrel-cage motor, but a _____ with windings brought out via slip rings and brushes.

6. The speed of wound-rotor induction motor is varied by

    a. Connecting variable resistance into the rotor circuit.

    b. Connecting variable resistance into the stator circuit.

    c. Applying variable voltage to the rotor circuit.

    d. Both a and c

    Answer _____

7. The lower the rotor resistance of a wound-rotor motor, the higher the speed of the motor. (True/False) _____

hands on PRACTICAL ASSIGNMENTS

1. The purpose of this assignment is to wire and operate different types of motor starters. Wire as many of the following motor-starting circuits as possible, using whatever type of motor control station is available to you. Complete a line diagram and a wiring diagram for each circuit and have them checked by the instructor. Wire each circuit in a neat and professional manner and have the installation evaluated by the instructor.

   - Single-phase, fractional-horsepower manual starter control of an AC motor.
   - Three-phase manual control of an AC motor.
   - Three-phase magnetically operated across-the-line starter with a single start-stop/station.
   - Three-phase magnetically operated across-the-line starter with two start/stop stations.
   - Reduced-voltage primary-resistance starter.
   - Reduced-voltage autotransformer starter.
   - Reduced-voltage wye-delta starter.
   - Reduced-voltage part-winding starter.
   - Solid-state soft-start starter.
   - DC across-the-line starter.

2. The purpose of this assignment is to wire and operate different types of motor reversing and jogging circuits. Wire as many of the following motor-reversing and jogging circuits as possible, using whatever type of motor control station is available to you. Complete a line diagram and a wiring diagram for each circuit and have them checked by the instructor. Wire each circuit in a neat and professional manner and have the installation evaluated by the instructor.

   - Across-the-line reversing starter with mechanical interlocking.
   - Across-the-line reversing starter with mechanical and electrical auxiliary contact interlocking.
   - Across-the-line reversing starter with mechanical and electrical pushbutton interlocking.
   - Three-phase across-the-line starter with jog, start, and stop control

3. The purpose of this assignment is to wire and operate different types of motor-stopping circuits. Wire as many of the following motor-stopping circuits as possible, using whatever type of motor control station is available to you. Complete a line diagram and a wiring diagram for each circuit and have them checked by the instructor. Wire each circuit in a neat and professional manner and have the installation evaluated by the instructor.

   - Plugging of a three-phase induction motor using a reversing starter and zero-speed switch.
   - Antiplugging of a three-phase induction motor using a reversing starter and zero-speed switch.
   - Dynamic braking of a DC motor.
   - DC injection braking of an induction motor.
   - Electromechanical friction braking of a motor.

4. The purpose of this assignment is to wire and operate different types of motor speed control circuits. Wire as many of the following motor speed control circuits as possible, using whatever type of motor control station is available to you. Complete a line diagram and a wiring diagram for each circuit and have them checked by the instructor. Wire each circuit in a neat and professional manner and have the installation evaluated by the instructor.

   - Multispeed motor starter.
   - Wound-rotor motor controller

"THE CONSTRUCTOR" CIRCUIT ANALYSIS ASSIGNMENTS

1. The purpose of this assignment is to observe the operation of a magnetic across-the-line starter, shown wired for two start/stop stations. Download circuit file Chapter 8 Simulation Assignment 01 (Figure 8-13, p. 166) from the assignment diagrams folder. Click on the Main Power Switch (ON/OFF) located on the left side of the toolbar. Left-click on the circuit breaker to close it and on the motor start and stop push buttons to operate them.

a. How are the two start push buttons connected relative to each other?

_____

b. How are the two stop push buttons connected relative to each other?

_____

c. Right-click on one of the stop buttons and lock it in the open position. What happens when the circuit is operated? Why?

_____

_____

d. Connect the ground fault to the coil side of the start push button. What happens when the circuit is operated? Why?

_____

_____

Figure 8-13 Across-the-line starter circuit with two start/stop stations.

2. The purpose of this assignment is to observe the automatic sequential starting of two motors at full line voltage. In order to reduce the amount of starting current, the circuit has been designed so that there will be a short time-delay period between the starting of motor 1 and motor 2. Download circuit file Chapter 8 Simulation Assignment 02 (Figure 8-14, p. 168) from the assignment diagrams folder. Click on the Main Power Switch (ON/OFF) located on the left side of the toolbar. Left-click on the circuit breaker to close it and on the motor start and stop button to operate them.

   a. What is time-delay period between the starting of the two motors?

   _____

   b. With both motors operating, left-click on motor OL-2 to lock it open. Describe how this affects the operation of the circuit while running and on restarting.

   _____

   _____

   c. With both motors operating, left-click on motor OL-1 to lock it open. Describe how this affects the operation of the circuit while running and on restarting. _____

   _____

   d. Left-click on the TR coil to simulate an open in the coil. What happens? Why? _____

   _____

   e. Right-click on starter coil M1 to simulate a shorted coil fault. What happens? Why? _____

   _____

Figure 8-14 Automatic sequential starting of two motors.

3. The purpose of this assignment is to observe the automatic sequential stopping of two motors. Download circuit file Chapter 8 Simulation Assignment 03 (Figure 8-15, p. 169) from the assignment diagrams folder. Click on the Main Power Switch (ON/OFF) located on the left side of the toolbar. Left-click on the circuit breaker to close it and on the motor start and stop buttons to operate them.

   a. Run the circuit simulation and summarize the sequence of operations.

   _____

   _____

   _____

   _____

   b. Right-click the normally closed M-1 contact to lock it in the closed position. In what way does this change the operation of the circuit?

   _____

   _____

Figure 8-15 Automatic sequential stopping of two motors.

4. The purpose of this assignment is to observe the operation of a primary resistance AC starter circuit. Download circuit file Chapter 8 Simulation Assignment 04 (Figure 8-16, p. 170) from the assignment diagrams folder. Click on the Main Power Switch (ON/OFF) located on the left side of the toolbar. Left-click on the circuit breaker to close it and on the motor start and stop buttons to operate them.

a. What is the preset value of the time-delay period?

_____

b. What occurs in the power circuit when the run contactor coil is energized?

_____

_____

c. Left-click on the starting resistors to simulate an open fault in each of them. In what way does this change the operation of the circuit?

_____

_____

_____

Figure 8-16   Primary resistance AC starter circuit.

5.   The purpose of this assignment is to observe the operation of a wye-delta starter circuit. Download circuit file Chapter 8 Simulation Assignment 05 (Figure 8-17, p. 171) from the assignment diagrams folder. Click on the Main Power Switch (ON/OFF) located on the left side of the toolbar. Left-click on the circuit breaker to close it and on the motor start and stop buttons to operate them.

a.   Which coil(s) is/are energized when the motor is first started?

_____

b.   Which coil(s) is/are energized after the time-delay period has elapsed?

_____

c.   Left-click on relay coil TR to fault it open. In what way does this change the operation of the motor?

_____

Figure 8-17 Wye-delta starter circuit.

6. The purpose of this assignment is to observe the operation of a reversing starter wired with auxiliary contactor interlocking. Download circuit file Chapter 8 Simulation Assignment 06 (Figure 8-18, p. 172) from the assignment diagrams folder. Click on the Main Power Switch (ON/OFF) located on the left side of the toolbar. Left-click on the circuit breaker to close it and on the motor forward and reverse buttons to operate them.

   a. What prevents the reverse contactor coil from being energized when the motor is operating in the forward direction?

   _____

   _____

   b. What prevents the forward contactor coil from being energized when the motor is operating in the reverse direction?

   _____

   _____

c. With the motor operating in either direction, how is the direction of rotation reversed?

_____

_____

d. Click on the forward NC contactor auxiliary contact and lock it in the open state. In what way is the operation of the circuit altered?

_____

_____

e. Click on the reverse N.O. contactor auxiliary contact and lock it in the closed state. In what way is the operation of the circuit altered?

_____

_____

Figure 8-18 Reversing starter with auxiliary contactor interlocking.

7. The purpose of this assignment is to observe the operation of a reversing starter wired with both electrical auxiliary contact and push button interlocking. Download circuit file Chapter 8 Simulation Assignment 07 (Figure 8-19) from the assignment diagrams folder. Click on the Main Power Switch (ON/OFF) located on the left side of the toolbar. Left-click on the circuit breaker to close it and on the motor forward and reverse buttons to operate them.

    a. Click on the forward push button. What change of state of the button's contacts occurs?

    _____

    _____

    b. Click on the reverse push button. What change of state of the button's contacts occurs?

    _____

    _____

    c. With the motor operating in the forward direction, try reversing the direction of rotation without first pressing the stop push button. What happens? Why?

    _____

    _____

Figure 8-19 Reversing starter with both auxiliary contact and pushbutton interlocking.

8. The purpose of this assignment is to observe the operation of a reversing starter wired with limit switches connected to limit travel. Download circuit file Chapter 8 Simulation Assignment 08 (Figure 8-20) from the assignment diagrams folder. Click on the Main Power Switch (ON/OFF) located on the left side of the toolbar. Left-click on the circuit breaker to close it and on the directional buttons and limit switches to operate them.

    a. Operate the circuit with the contacts of both limit switches closed. Why is travel not restricted in either direction?

    _____

    b. Operate the circuit with the contact of the forward limit switch open and the reverse limit switch contact closed. In what way is travel restricted?

    _____

    c. Operate the circuit with the contact of the reverse limit switch open and the forward limit switch contact closed. In what way is travel restricted?

    _____

Figure 8-20 Reversing starter wired with limit switches connected to limit travel.

9. The purpose of this assignment is to observe the operation of a reversing starter wired for a reciprocating machine process. Download circuit file Chapter 8

Simulation Assignment 09 (Figure 8-21) from the assignment diagrams folder. Click on the Main Power Switch (ON/OFF) located on the left side of the toolbar. Left-click on the circuit breaker to close it and on the buttons and limit switches to operate them.

a. With the motor traveling in the forward direction, what happens when the forward limit switch is actuated? _____

b. With the motor traveling in the reverse direction, what happens when the reverse limit switch is actuated? _____

c. Assume that the forward limit switch sticks and fails to return to its normal deactivated position once the motor starts operating in the reverse direction. Simulate this scenario. In what way is the operation of the circuit altered as a result of this fault condition?

_____

_____

Figure 8-21   Reciprocating machine process.

175

10. The purpose of this assignment is to observe the operation of a pushbutton jog circuit. Download circuit file Chapter 8 Simulation Assignment 10 (Figure 8-22) from the assignment diagrams folder. Click on the Main Power Switch (ON/OFF) located on the left side of the toolbar. Left-click on the circuit breaker to close it and on the buttons to operate them.

    a. With the motor operating in the run mode, what happens when the jog push button is actuated? _____

    b. Assume the N.O. make jog contact becomes locked in the open state. Simulate this scenario. In what way is the operation of the circuit altered as a result of this fault condition?

    _____

    _____

    c. Assume that on release of the jog push button, the NC pushbutton contacts reclose before the starter maintaining contact M opens. What potential hazard operating condition would be created?

    _____

Figure 8-22   Pushbutton jog circuit.

11. The purpose of this assignment is to observe the operation of a control relay jog circuit. Download circuit file Chapter 8 Simulation Assignment 11 (Figure 8-23) from the assignment diagrams folder. Click on the Main Power Switch (ON/OFF) located on the left side of the toolbar. Left-click on the circuit breaker to close it and on the buttons to operate them.

a. Which coil is directly energized by closing the start push button?

_____

b. With the motor operating in the run mode, what happens when the jog button is pressed? Why?

_____

_____

c. Assume the CR coil is faulted open. Simulate this scenario. In what way is the operation of the circuit altered as a result of this fault condition?

_____

d. Assume the M coil is faulted open. Simulate this scenario. In what way is the operation of the circuit altered as a result of this fault condition?

_____

Figure 8-23   Control relay jog circuit.

12. The purpose of this assignment is to observe the operation of a start/stop/selector jog circuit. Download circuit file Chapter 8 Simulation Assignment 12 (Figure 8-24) from the assignment diagrams folder. Click on the Main Power Switch (ON/OFF) located on the left side of the toolbar. Left-click on the circuit breaker to close it and on the buttons to operate them.

   a. What state must the run/jog switch be in to operate the motor in the jog mode? _____

   b. With the motor operating in the run mode, what happens when the run/jog switch is opened? Why? _____

   c. Assume the stop push button is locked in the open position. Simulate this scenario. In what way is the operation of the circuit altered as a result of this fault condition?

   _____

Figure 8-24   Start/stop/selector jog circuit.

13. The purpose of this assignment is to observe the operation of an antiplugging protection circuit. Download circuit file Chapter 8 Simulation Assignment 13 (Figure 8-25) from the assignment diagrams folder. Click on the Main Power Switch (ON/OFF) located on the left side of the toolbar. Left-click on the circuit breaker to close it and on the switch buttons to operate them. Activate the F and R zero-speed switch contacts with respect to the speed and direction rotation of the motor.

a. Immediately after the forward push button is pressed, with the motor operating, what state (open or closed) should the F and R zero-switch contacts be in? Simulate this scenario.

---

b. With the motor running in the forward direction what state should the F and R zero-switch contacts be in immediately after pressing the stop push button? Simulate this scenario.

---

c. At what point after this will pressing the reverse push button allow the reverse contactor to be energized? Simulate this scenario.

Figure 8-25 Antiplugging protection circuit.

14. The purpose of this assignment is to observe the operation of a dynamic braking circuit for a DC motor. Download circuit file Chapter Simulation Assignment 14 (Figure 8-26) from the assignment diagrams folder. Click on the Main Power Switch (ON/OFF) located on the left side of the toolbar. Left-click on the circuit breaker to close it and on the start and stop buttons to operate them.

   a. With coil M deenergized, how is the dynamic braking resistor connected relative to the motor armature? _____

   b. When is the only time that current flows through the dynamic braking resistor?

      _____

   c. Assume the dynamic braking resistor is faulted shorted. How would this affect the period of time for the motor to stop? Why?

      _____

      _____

Figure 8-26 Dynamic braking circuit for a DC motor.

15. The purpose of this assignment is to observe the operation of an AC motor DC injection braking circuit. Download circuit file Chapter 8 Simulation Assignment 15 (Figure 8-27, p. 182) from the assignment diagrams folder. Click on the Main Power Switch (ON/OFF) located on the left side of the toolbar. Left-click on the circuit breaker to close it and on the start and stop buttons to operate them.

    a. What is the function of the rectifier? _____

    b. What type of pushbutton contact arrangement is used for the start button?

        _____

    c. What is the state of each of the coils immediately after the start push button has been pressed? _____

    d. What is the state of each of the coils immediately after the stop push button has been pressed? _____

    e. What is the state of each of the coils after the injection braking has been applied?

        _____

    f. Assume that one lead on the secondary of the braking transformer is moved to the center tap terminal. Would this increase or decrease the amount of time required to bring the motor to a stop? Why?

        _____

Figure 8-27 AC motor DC injection braking circuit.

9 Motor Control Electronics

PART 1   Semiconductor Diodes

Place the answers in the space provided.

1. A diode consists of _____-type and _____-type
    semiconductor materials formed on a single component.

2. The most important operating characteristic of a diode is that it allows current to
    flow through it in both directions. (True/False) _____

3. Identify the operating modes of the diodes shown in Figure 9-1.

    a. _____ bias   b. _____ bias

Figure 9-1   Diode for Question 3.

4. Identify the leads of the diode shown in Figure 9-2.

    a. _____   b. _____

Figure 9-2   Diode for Question 4.

5. When voltage is applied across a diode in such a way that the diode allows current to flow, the diode is said to be

    a. Short-circuited

    b. Open-circuited

    c. Forward-biased

    d. Reverse-biased

    Answer _____

6. Connect the devices shown in Figure 9-3 in series so that the lamp turns on when the switch is closed.

Figure 9-3   Devices for Question 6.

7. Diode rectifiers are used in changing _____ to _____.

8. Answer the following with reference to the rectifier circuit shown in Figure 9-4.

    a. This circuit would be classified as a _____-wave rectifier.

    b. During the positive half-cycle of the AC input, the diode acts as a _____ switch.

    c. During the negative half-cycle of the AC input the diode acts as a _____ switch.

    d. The circuit produces a continuous DC voltage across the load. (True/False) _____

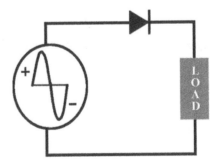

Figure 9-4   Rectifier circuit for Question 8.

9. An ohmmeter test of a diode shows a low resistance reading in both directions. This is an indication that the diode is

a. Operating normally

b. Faulted shorted

c. Faulted open

d. Faulted grounded

Answer _____

10. Answer the following with reference to the diode control relay circuit shown in Figure 9-5.

a. This circuit is an application of a diode-_____ circuit.

b. The reason the diode is required is because the CR relay coil is mainly a resistive load. (True/False) _____

c. Closing the switch results in current flowing through the _____ but not the _____.

d. When the switch is opened, the _____ voltage of the coil bleeds off through the diode.

e. Reversing the diode lead connections will result in a short circuit when the switch is closed. (True/False) _____

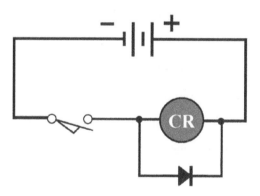

Figure 9-5   Relay circuit for Question 10.

11. Answer the following with reference to the rectifier circuit shown in Figure 9-6.

    a. This circuit would be classified as a _____-wave rectifier.

    b. On the positive half-cycle, diodes _____ and _____ will conduct.

    c. On the negative half-cycle, diodes _____ and _____ will conduct.

    d. The shape of the voltage waveform across the load will be similar to the AC input sine wave with the_____ half of the waveform inverted.

    e. The DC polarity of point A relative to point B will be _____.

    f. Reversing the two leads of the AC input voltage will result in a reversal of the DC polarity across the load. (True/False) _____

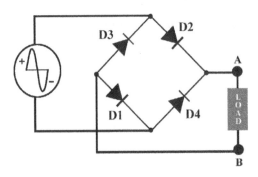

Figure 9-6   Rectifier circuit for Question 11.

12. The purpose of a power supply filter is to

    a. Change alternating current to direct current

    b. Change direct current to alternating current

    c. Smooth out the DC voltage from the diode rectifier circuit

    d. Produce a cleaner form of the original AC input voltage

    Answer _____

13. The DC output voltage of a diode rectifier circuit can be smoothed out by means of a

    a. Resistor connected in series with the output

    b. Resistor connected in parallel with the output

    c. Capacitor connected in series with the output

    d. Capacitor connected in parallel with the output

    Answer _____

186

14. Answer the following with reference to the rectifier circuit shown in Figure 9-7.

    a. This circuit would be classified as a _____ full-wave bridge rectifier.

    b. Compared to single-phase rectifiers, this circuit requires much more filtering to obtain a low-ripple DC output. (True/False) _____

    c. When this rectifier is constructed in modular form, it comes equipped with _____ AC input leads and _____ DC output leads.

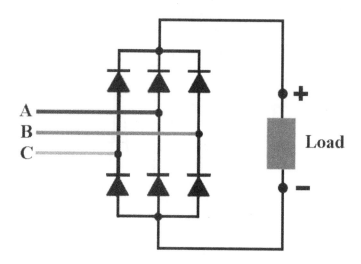

Figure 9-7 Rectifier circuit for Question 14.

15. A zener diode

    a. Allows current to flow in the forward direction

    b. Allows current to flow in the reverse direction when the voltage is above a certain value

    c. Is typically used to regulate voltage in electric circuits

    d. All of these

    Answer _____

16. Answer the following with reference to the circuit shown in Figure 9-8.

    a. The function of this circuit is to suppress damaging AC _____ voltages.

    b. The varistor module is made up of two _____ diodes connected back to back.

    c. Each diode acts as an open circuit until the reverse bias voltage across it exceeds its rated value. (True/False) _____

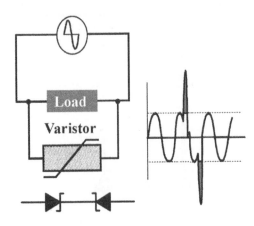

Figure 9-8 Circuit for Question 16.

17. A light-emitting diode (LED) will

    a. Conduct a current in one direction only

    b. Give off light when current flows through it

    c. Consume less energy than a filament-type light bulb

    d. All of these

    Answer _____

18. Answer the following with reference to the optocoupler circuit shown in Figure 9-9.

    a. The diode in the input circuit is a _____ and is connected in _____ bias.

    b. The diode in the output circuit is a _____ and is connected in _____ bias.

    c. The only thing connecting the two circuits is _____.

    d. When the push button is closed, _____ enters the photodiode so its resistance decreases, switching on current to the load.

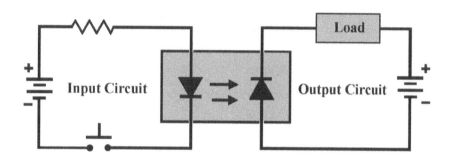

Figure 9-9   Optocoupler circuit for Question 18.

PART 2   Transistors

1. A transistor is used to amplify a signal or switch a circuit on and off. (True/False) _____

2. The two most common types of transistors are the _____ transistor and the _____-_____ transistor.

3. The three leads of a bipolar transistor are the

    a. Anode, base, and cathode

    b. Emitter, gate, and anode

    c. Base, collector, and emitter

    d. Grid, anode, and cathode

    Answer _____

4. In a bipolar transistor circuit, the collector current is typically much higher than the base current. (True/False) _____

5. The PNP transistor is the complement of the NPN transistor. This means that

   a. The two operate on different principles.

   b. One can be replaced by the other in a given circuit.

   c. The two are electrically similar except that opposite current and voltages are involved.

   d. One is designed to operate from an AC source and the other from a DC source.

   Answer _____

6. Answer the following with reference to the transistor amplifying circuit shown in Figure 9-10.

   a. The bipolar transistor would be classified as an NPN type. (True/False) _____

   b. The base current varies in direct proportion with the amount of light shining on the photovoltaic sensor. (True/False) _____

   c. With no light shining on the photovoltaic sensor, the voltage and current generated by it would be _____.

   d. As the base current increases, the collector current decreases. (True/False) _____

   e. The current amplification factor or gain for this circuit would be

   _____.

Figure 9-10   Transistor amplifying circuit for Question 6.

7. Answer the following with reference to the transistor switching circuit shown in Figure 9-11.

    a. For the circuit shown the 230-V load is operated directly by the transistor. (True/False) _____

    b. When the proximity sensor switch is open, the transistor conducts to energize the relay coil. (True/False) _____

    c. The relay contacts used to control the load would be classified as being normally _____.

    d. When the transistor is in the off state, the voltage drop across the collector and emitter will be approximately_____ V and the voltage across the relay coil approximately _____ V.

    e. The diode used in this circuit prevents the induced coil voltage at turn-off from damaging the transistor. (True/False) _____

Figure 9-11   Transistor switching circuit for Question 7.

8. A Darlington transistor combines _____ into a single device.

    a. Two transistors

    b. A transistor and relay

    c. A transistor and diode

    d. A transistor and LED

    Answer _____

9. Identify the labeled components of the phototransistor optical isolator circuit shown in Figure 9-12.

    a. _____

    b. _____

    c. _____

    d. _____

Figure 9-12 Optical isolator circuit for Question 9.

10. Compared to a bipolar transistor, the field-effect transistor (FET)

    a. Has a lower switching speed

    b. Requires less drive power

    c. Is less likely to be damaged by electrostatic discharge

    d. Both a and b

    Answer _____

11. Identify the three leads of the junction field-effect transistor (JFET) shown in Figure 9-13.

    a. _____

    b. _____

    c. _____

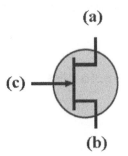

Figure 9-13 Junction field-effect transistor for Question 11.

12. All field-effect transistors are unipolar in that their working current flows through only one type of semiconductor material.
(True/False) _____

13. All field-effect transistors use basically no input power, as the output current flow is controlled by an electric field rather than an electric current.
(True/False) _____

14. Answer the following with reference to the JFET circuit shown in Figure 9-14.

    a. When the _____ voltage is zero, maximum current flows between the source and drain.

    b. A negative voltage applied to the gate decreases the channel resistance.
    (True/False) _____

    c. Passage of current between the source and drain is essentially blocked when the gate voltage reaches the pinch-off point. (True/False) _____

Figure 9-14   JFET circuit for Question 14.

15. The gate of a JFET consists of a _____-biased junction, whereas the gate of a MOSFET consists of a metal electrode _____ from the channel.

16. Identify the modes of N-channel MOSFETs represented by the symbols shown in Figure 9-15.

a. _____ mode   b. _____ mode

Figure 9-15   N-channel MOSFETs for Question 16.

17. Which of the following field-effect transistors is normally off, in that if a voltage is connected to the drain and source and no voltage is connected to the gate, there will be no current flow through the device?

a. JFET

b. Depletion-mode MOSFET

c. Enhancement-mode MOSFET

d. Both a and b

Answer _____

18. Answer the following with reference to the MOSFET off-delay timer circuit shown in Figure 9-16 (p. 195).

a. With the external switch initially open, the relay coil will be energized. (True/False) _____

b. Closing the external switch charges the capacitor and triggers the MOSFET into conduction. (True/False) _____

c. When the external switch is closed, the timing period begins. (True/False) _____

d. When the external switch is opened, the charged capacitor begins to discharge its stored energy through _____ and _____.

e. If the resistance of the variable resistor R2 is increased, the length of the time-delay period will decrease. (True/False) _____

Figure 9-16 MOSFET off-delay timer for Question 18.

19. Compared to mechanical switching circuits, transistor switching circuits

    a. Require less driving power to operate

    b. Can be operated on and off at a higher rate of speed

    c. Have no moving parts to wear out

    d. All of these

    Answer _____

20. Answer the following with reference to the power MOSFET motor control circuit shown in Figure 9-17 (p. 196).

    a. In this application, the MOSFET is used as part of a chopper circuit to switch voltage to the _____ on and off very rapidly.

    b. By switching the voltage, an average voltage somewhere between 0 and 100 percent is produced. (True/False) _____

c. When operated with a 50 percent duty cycle, the motor will run slower than when operated with an 80 percent duty cycle. (True/False) _____

d. The _____-biased diode provides a discharge path when the motor armature is switched off.

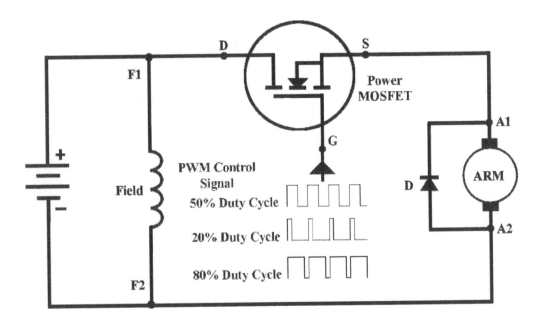

Figure 9-17   MOSFET motor control circuit for Question 20.

21. Identify the leads of the insulated-gate bipolar transistor (IGBT) shown in Figure 9-18.

   a. _____

   b. _____

   c. _____

Figure 9-18   IGBT for Question 21.

22. The insulated gate bipolar transistor is a cross between a _____ and a _____ transistor.

23. IGBTs have a lower on-state power loss than MOSFETs. (True/False) _____

24. IGBTs have faster switching speeds than BJTs. (True/False) _____

25. Answer the following with reference to the IGBT motor drive circuit shown in Figure 9-19.

   a. The input power from the line is _____-phase AC.

   b. The six-diode bridge configuration converts the _____ power to _____ power.

   c. The inductor and capacitor of the DC bus section acts to _____ out any AC components of the DC waveform.

   d. The six-IGBT bridge configuration inverts the _____ power back to _____ power.

   e. Control circuit switching of the IGBT devices delivers a variable _____ and _____ output to the motor.

Figure 9-19   IGBT motor drive circuit for Question 25.

PART 3   Thyristors

1. Thyristors have only two states—_____ and _____.

2. Thyristors are mainly used where low currents and voltages are involved. (True/False) _____

3. Identify the leads of the silicon controlled rectifier (SCR) shown in Figure 9-20.

   a. _____

   b. _____

   c. _____

Figure 9-20

SCR for Question 3.

197

4. High-current SCRs have provisions for some type of heat _____ to dissipate the heat.

5. How is a silicon-controlled rectifier similar to a diode rectifier?

   a. Both conduct current in one direction only.

   b. Both can be classified as thyristors

   c. Both are constructed using only one PN junction

   d. All of these

   Answer _____

6. Answer the following with reference to the direct-current SCR switching circuit shown in Figure 9-21.

   a. Momentarily pressing pushbutton _____ switches the SCR into conduction.

   b. Once switched on, the SCR conducts like a _____.

   c. Once conducting, the only way to turn the SCR off is to reduce the _____-to-_____ current to zero.

   d. Reversing the polarity of the DC supply voltage will reverse which push buttons turn the lamp on and off. (True/False) _____

Figure 9-21 Direct current SCR switching circuit for Question 6.

7. Answer the following with reference to the alternating current SCR switching circuit shown in Figure 9-22.

    a. Momentarily closing PB will switch the lamp on, and it will remain on after the push button returns to its normally open position.
(True/False) _____

    b. The SCR is automatically shut off during each _____ half-cycle because it is _____-biased.

    c. Holding PB closed will produce a _____-wave pulsating direct current through the lamp.

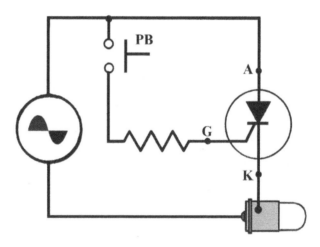

Figure 9-22   Alternating current SCR switching circuit for Question 7.

8. When an SCR is operated from an AC source

    a. The gate can be used to switch the current on and off at all times

    b. Once switched on by the gate, the current cannot be switched off by the gate

    c. The output is always direct current

    d. Both b and c

    Answer _____

9. Answer the following with reference to the alternating current SCR circuit shown in Figure 9-23 (p. 200) used to vary the amount of power delivered to a load.

    a. The load voltage is varied by using phase shift control.
(True/False) _____

    b. A pulse trigger is applied to the _____ lead at the instant that the SCR is required to turn on.

c. During the positive half of the AC input waveform, SCR-_____
and SCR-_____ can be triggered into conduction.

d. During the negative half of the AC input waveform, SCR-_____
and SCR-_____ can be triggered into conduction.

e. Power is regulated by advancing or delaying the point at which each pair of
SCRs is turned on within each half cycle. (True/False) _____

f. Applying the trigger pulse at the midpoint of the half-cycle will result in the
delivery of maximum power to the load. (True/False) _____

g. The current through the load reverses as the AC source voltage alternates
from one half-cycle to the other. (True/False) _____

Figure 9-23 Alternating current SCR circuit for Question 9.

10. Answer the following with reference to the SCR solid-state reduced-voltage
starter circuit shown in Figure 9-24 (p. 201).

a. In this application, a controlled AC output is obtained by connecting two
SCRs in _____-parallel.

b. When the motor is first started, contacts _____ close to apply a
reduced voltage to the motor.

c. Controlled triggering of the SCRs allows portions of the AC input waveform
to be applied to the motor. (True/False) _____

d. The current transformers are used to monitor motor speed.
(True/False) _____

e. After the starting period, contacts _____ open and contacts
_____ connect the motor directly across the line.

Figure 9-24 SCR solid-state reduced-voltage starter circuit for Question 10.

11. SCRs usually fail shorted as indicated by an ohmmeter resistance reading from anode to cathode that is

    a. High in one direction and low in the other

    b. High in both directions

    c. Near infinity in both directions

    d. Near zero in both directions

    Answer _____

12. The physical size of an SCR has no relationship to its current rating.

    (True/False) _____

13. Identify the leads of the triac shown in Figure 9-25.

    a. _____

    b. _____

    c. _____

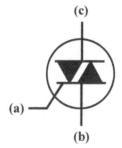

Figure 9-25 Triac for Question 13.

14. Current flow between the main terminals of a triac is

    a. In one direction only

    b. In either direction

    c. Constant at all times

    d. Less than the gate current flow value

    Answer _____

15. Answer the following with reference to the triac circuit shown in Figure 9-26.

    a. This circuit is used to switch DC loads. (True/False) _____

    b. When the switch is closed and then opened, the triac will conduct and remain conducting. (True/False) _____

    c. The function of the resistor is to limit the flow in the gate circuit to a small control current value. (True/False) _____

    d. The current rating of the switch can be much lower than that of the load. (True/False) _____

Figure 9-26 Triac circuit for Question 15.

16. A triac can serve as the switching device for a programmable logic controller AC output module. (True/False) _____

17. Answer the following with reference to the triac variable AC control circuit shown in Figure 9-27 (p. 203).

    a. The control logic controls the _____ on the AC waveform at which the triac is switched on.

    b. The output waveform is still AC, but the _____ current value is adjustable.

c. This circuit could be used to vary the speed of a single-phase universal motor. (True/False) _____

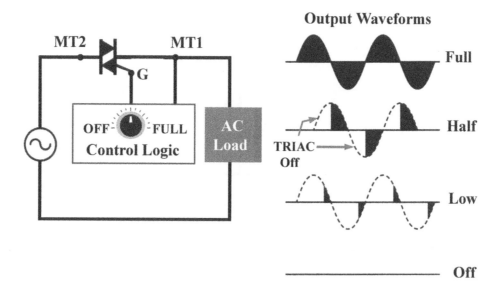

Figure 9-27 Triac AC control circuit for Question 17.

18. A diac is designed to conduct current in

a. Both directions like a resistor

b. The forward-bias direction only

c. The reverse-bias direction only

d. Either direction when its rated brakeover voltage is exceeded

Answer _____

19. Answer the following with reference to the lamp dimmer circuit shown in Figure 9-28 (p. 204).

a. Maximum lamp brightness is achieved when the variable resistor is set to its maximum resistance value. (True/False) _____

b. The diac conducts at all times that the capacitor is charging. (True/False) _____

c. If the triac is faulted shorted, the lamp will be at maximum brightness at all times. (True/False) _____

Figure 9-28 Lamp dimmer circuit for Question 19.

PART 4 Integrated Circuits (ICs)

1. An integrated circuit (IC) is a

    a. Number of transistors mounted within a silicon chip

    b. Miniaturized electronic circuit fabricated into a single silicon chip

    c. Series of discrete components wired together

    d. Series of separately produced components wired within a printed circuit board

    Answer _____

2. ICs are favored for high-voltage and high-current applications.
    (True/False) _____

3. Digital ICs operate with on/off _____-type signals.

4. Analog ICs contain _____-type circuitry.

5. Identify the two pin numbers of the IC shown in
    Figure 9-29.

    a. _____ b. _____

(b)

Figure 9-29 IC for Question 5.   (a)

6. An operational amplifier (op-amp) would be classified as

   a. An analog IC

   b. A digital IC

   c. A MOS IC

   d. A TTL IC

   Answer _____

7. Answer the following with reference to the op-amp voltage-amplifier circuit shown in Figure 9-30.

   a. Resistor _____ is called the feedback resistor.

   b. The voltage gain of this circuit would be _____.

   c. When the peak-to-peak value of the input signal is 100 mV, the peak-to-peak value of the output signal would be _____.

   d. The output signal is in phase with the input signal.
      (True/False) _____

Figure 9-30 Op-amp circuit for Question 7.

8. Answer the following with reference to the op-amp voltage comparator circuit shown in Figure 9-31 (p. 206).

   a. The output is dependent on the difference in value between voltages V1 and V2. (True/False) _____

   b. When the light-dependent resistor (LDR) is not illuminated with light, its resistance is very high. (True/False) _____

c. When sufficient light shines on the LDR, voltage V1 will be greater than V2, resulting in an output from the op-amp activating the relay to switch the connected load. (True/False) _____

Figure 9-31 Op-amp voltage comparator circuit for Question 8.

9. Answer the following with reference to the 555 IC interval timer circuit shown in Figure 9-32 (p. 207).
    a. The length of the time interval is determined by the value of timing components _____ and _____.
    b. When the switch is opened, the capacitor is held charged. (True/False) _____
    c. When the switch is closed the capacitor starts to charge. (True/False) _____
    d. When the switch is closed, the LED is immediately switched on. (True/False) _____
    e. When the switch is opened, the LED is immediately switched off. (True/False) _____
    f. With the switch kept closed, the LED remains on until the charge on the capacitor reaches a certain level. (True/False) _____

Figure 9-32

IC interval timer circuit for
Question 9.

10. Answer the following with reference to the 555 DC motor speed controller circuit
shown in Figure 9-33.

   a. The type of speed control used in this circuit is known as _____

      _____   _____.

   b. The voltage applied to the motor armature is switched on and off by
      transistor Q2. (True/False) _____

   c. A duty cycle of 50 percent will result in a greater armature voltage and speed
      than one of 10 percent. (True/False) _____

   d. Potentiometer R1 controls the length of time the _____ of the
      timer will be turned on.

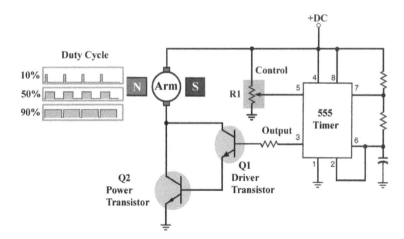

Figure 9-33

Speed control circuit
for Question 10.

11. A microcontroller is an integrated circuit that functions as a complete computer on a chip. (True/False) _____

12. An _____ microcontroller is one that is physically built into the device it controls.

13. Electrostatic discharge (ESD) is the release of static electricity when two objects come into contact. (True/False) _____

14. The resultant _____ flow from ESD can generate damaging heat within an IC.

15. Always transport and store sensitive ICs and control boards in _____ packaging.

16. Logic is the ability to make decisions based on one or more different factors before an action is taken. (True/False) _____

17. The only two states in digital logic circuits are _____ and

_____.

18. Digital circuits are constructed from electronic circuits known as logic

_____.

19. Identify the type of logic gate function represented by the symbols shown in Figure 9-34.

a. _____   b. _____   c. _____

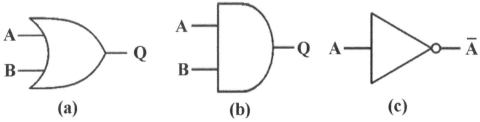

Figure 9-34 Symbols for Question 19.

20. A logic gate _____ table shows the output for each possible input.

21. The table shown in Figure 9-35
    summarizes the operation of a two-input
    a. AND gate
    b. OR gate
    c. NOT gate
    d. NAND gate
    Answer _____

Input A	Input B	Output Q
0	0	0
0	1	1
1	0	1
1	1	1

Figure 9-35 Table for Question 21.

22. The table shown in Figure 9-36
    summarizes the operation of a two-input
    a. AND gate
    b. OR gate
    c. NOT gate
    d. NAND gate
    Answer _____

Input A	Input B	Output Q
0	0	1
0	1	1
1	0	1
1	1	0

Figure 9-36 Table for Question 22.

23. Identify the type of gate logic function utilized in the hard-wired control circuit
    shown in Figure 9-37.
    Answer _____

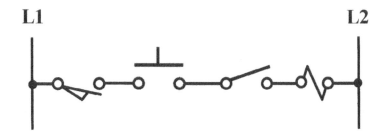

Figure 9-37 Control circuit for Question 23.

24. Identify the type of gate logic function utilized in the hard-wired control circuit shown in Figure 9-38.

Answer _____

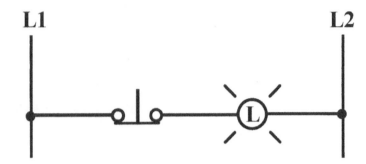

Figure 9-38 Control circuit for Question 24.

25. Identify the type of gate logic function utilized in the programmable logic controller (PLC) circuit shown in Figure 9-39.

Answer _____

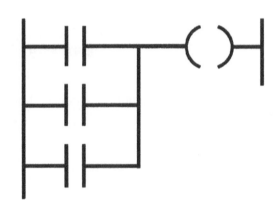

Figure 9-39 PLC circuit for Question 25.

10 Adjustable-Speed Drives and PLC Installations

PART 1   AC Motor Drive Fundamentals

Place the answers in the space provided.

1. An adjustable-speed drive may be used to control _____ of a motor

    a. Speed and torque

    b. Acceleration and deceleration

    c. Direction of rotation

    d. All of these

    Answer _____

2. The preferred method of speed control for squirrel-cage induction motors is to alter the _____ of the supply voltage.

3. Identify the blocks of the block diagram of the variable-frequency drive (VFD) shown in Figure 10-1.

    a. _____   b. _____

    c. _____   d. _____

Figure 10-1   Variable-frequency drive for Question 3.

4. A VFD can be used to operate a three-phase motor from a single-phase AC supply. (True/False) _____

5. Complete the wiring for the three-phase VFD converter circuit shown in Figure 10-2.

Figure 10-2 Three-phase VFD converter for Question 5.

6. The DC bus circuit of a VFD acts to _____ out any AC components of the DC waveform.

7. The DC bus voltage is calculated by taking the RMS value of the line voltage and multiplying it by
    a. 0.707
    b. 0.637
    c. 1.414
    d. 1.707
    Answer _____

8. The inverter circuit of a VFD acts to switch the DC bus voltage on and off at specific intervals. (True/False) _____

9. a. Complete the wiring for the VFD control logic and inverter circuit shown in Figure 10-3.
    b. The type of transistors used is _____-_____ bipolar transistor.
    c. The control logic and inverter control the output _____ and _____ to the motor.

Figure 10-3 VFD logic and inverter circuit for Question 9.

10. Answer the following with reference to the inverter voltage output waveforms shown in Figure 10-4.

   a. This inverter would be classified as a pulse-width modulated type. (True/False) _____

   b. The sine wave represents the _____ frequency and the pulses the _____ frequency.

   c. The pulses may be of different heights. (True/False) _____

   d. The output voltage is varied by changing the _____ and _____ of the switched pulses.

   e. The output _____ is adjusted by changing the switching cycle time.

   f. The fundamental frequency is a fixed frequency substantially higher than the switching frequency. (True/False) _____

   g. The whining sound associated with VFDs is generated by the fundamental frequency. (True/False) _____

   h. Higher carrier frequency allows a better approximation to the sinusoidal form of the output current. (True/False) _____

Figure 10-4 Waveforms for Question 10.

11. The _____-duty motor is designed for optimized performance to operate in conjunction with a VFD.

12. An open-loop VFD operates without benefit of any _____ signals.

13. With VFD speed control the voltage to the motor must be increased as the frequency is decreased to prevent overheating of the motor.
(True/False) _____

14. Answer the following with reference to the VFD voltage versus frequency (V/Hz) graph shown in Figure 10-5.

    a. The ratio between voltage and frequency does not change as the motor voltage is increased from 0 to 460 V. (True/False) _____

    b. When 230 V is applied to the motor, the frequency at which this voltage is applied will be approximately _____ Hz.

    c. As the voltage to the motor increases, the motor speed decreases. (True/False) _____

    d. As the voltage increases from zero to the base speed 460-V level, the _____ remains constant.

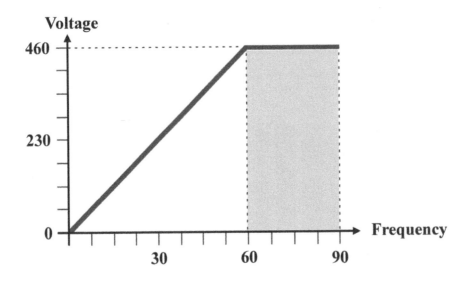

Figure 10-5 Voltage versus frequency graph for Question 14.

15. Which of the following VFD refinements compensates for power losses that occur at low speeds?

    a. Stability compensation

    b. Steady-state slip compensation

    c. Low-frequency voltage boost

    d. Both a and b

    Answer _____

214

16. The flux vector method of VFD speed control provides less precise control of motor speed and torque. (True/False) _____

17. Flux vector VFDs are classified as _____ types.

    a. Sensor

    b. Sensorless

    c. Magnetic

    d. Either a or b

    Answer _____

18. A motor-attached encoder device may be used to provide feedback on the motor's

    a. Rotor position

    b. Speed

    c. Direction

    d. All of these

    Answer _____

PART 2   VFD Installation and Programming Parameters

1. The load characteristics of the driven machinery are not of importance when you are selecting a variable-frequency drive for a particular application. (True/False) _____

2. A VFD reactor is basically an _____ installed on the input or output of the drive.

3. Use of which of the following will increase the distance that the motor can be located from the drive?

    a. Circuit breaker protection

    b. Fuse protection

    c. Load reactor

    d. Line reactor

    Answer _____

4. Which of the following should be taken into consideration in determining the location for a VFD?

   a. Distance between the motor and the drive

   b. Surrounding environment

   c. Working space to carry out maintenance

   d. All of these

   Answer _____

5. VFD _____ are designed to provide protection from the environment.

6. Answer the following with reference to the VFD operator interface shown in Figure 10-6.

   a. The three modes of operation are _____, _____, and

      _____.

   b. The interface provides a means for the operator to _____ and

      _____ the motor.

   c. List five motor typical operating parameters that may be viewed in real time on the monitor.

      (1) _____ (2) _____

      (3) _____ (4) _____

      (5) _____

Figure 10-6 VFD operator interface for Question 6.

7. Electromagnetic interference (EMI) is generated in a VDF as a result of the high switching _____ of the transistors.

8. Most manufacturers require that a shielded power cable be used for the electrical connection between the VFD and the motor. (True/False) _____

9. In addition to the safety aspect, proper grounding of a VFD system reduces electromagnetic interference. (True/False) _____

10. For the bypass contactor power circuit shown in Figure 10-7, when the motor is required to be directly operated from the line, the

    a. Isolation contacts are opened and the bypass contacts closed

    b. Isolation contacts are closed and the bypass contacts open

    c. Both the isolation and bypass contacts are opened

    d. Both the isolation and bypass contacts are closed

    Answer _____

Figure 10-7 Bypass contactor circuit for Question 10.

11. VFDs are not required to have a disconnecting means capable of disconnecting all power supplied to the drive. (True/False) _____

12. VFDs can operate as motor protection devices along with their role of motor speed controllers. (True/False) _____

13. Excessive energy is generated when the _____ drives the motor during deceleration.

14. Which of the following VFD braking systems redirects energy back to the AC supply?

    a. Dynamic braking

    b. Regenerative braking

    c. Friction braking

    d. Both a and b

    Answer _____

15. _____ is the ability of a VFD to increase or decrease the voltage and frequency to an AC motor gradually.

16. VFDs increase motor speed in steps. (True/False) _____

17. Variable-frequency drive inputs can be classified as being either
_____ or _____.

18. Which of the following devices would be considered to be digital in nature?
    a. Push button
    b. Selector switch
    c. Relay contact
    d. All of these
    Answer _____

19. A speed control potentiometer would be classified as an analog input.
    (True/False) _____

20. Which of the following motor nameplate data may be programmed into a VFD?
    a. Speed
    b. Full-load current
    c. Voltage
    d. All of these
    Answer _____

21. Derating a VFD means using a smaller than normal drive in the application.
    (True/False) _____

22. Derating may be required when a drive is installed at high altitude.
    (True/False) _____

23. Answer the following with reference to the VFD circuit shown in Figure 10-8 (p. 219).
    a. The drive would be classified as a voltage source inverter type.
       (True/False) _____
    b. This circuit uses _____ for the electronic switching devices.
    c. With this arrangement, the output from the converter is a _____ DC voltage.
    d. The inverter section takes six 60° steps to complete one 360° cycle.
       (True/False) _____
    e. One disadvantage of this type of drive is cogging that may occur at _____ speeds.

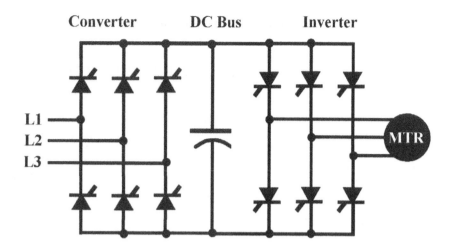

Figure 10-8 VFD circuit for Question 23.

24. The PID controller found with some VFDs automatically corrects the error between the actual motor speed and the desired setpoint speed by calculating the _____ between the two and then outputting a corrective action.

25. Tuning the PID controller involves _____ and _____ adjustments that result in a fast response with minimum overshoot.

26. Which of the following would *not* be considered a VFD program parameter?

    a. Minimum and maximum speeds

    b. Acceleration and deceleration rates

    c. Type of enclosure

    d. Carrier frequency

    Answer _____

27. Which of the following types of program parameters can be changed or adjusted while the drive is running or stopped?

    a. Processor firmware

    b. Tunable on the fly

    c. Configurable

    d. Read-only

    Answer _____

28. Examples of drive display parameters include _____, _____, _____, and _____.

29. Nonconfigurable alarms can be enabled or disabled by the operator.

    (True/False) _____

PART 3   DC Motor Drive Fundamentals

1. The speed of a DC motor is _____ proportional to the armature

    voltage and _____ proportional to the field current.

2. To change the direction of rotation of a DC motor, you

    a. Reverse the armature polarity

    b. Reverse the field polarity

    c. Reverse either the armature or field polarity

    d. Reverse both the armature and field polarity

    Answer _____

3. Answer the following with reference to the block diagram of the DC drive shown

    in Figure 10-9.

    a. This drive uses _____ semiconductor switching.

    b. The shunt field is supplied with a variable DC voltage.

        (True/False) _____

    c. The power converter block converts the supply voltage to an adjustable

        _____ _____ voltage.

    d. The current feedback signal gauges motor _____ while the

        voltage feedback signal gauges motor _____.

Figure 10-9   DC drive for Question 3.

4. Complete the wiring for the DC drive circuit shown in Figure 10-10.

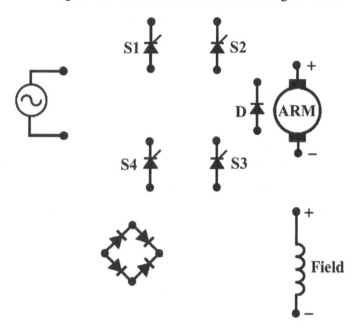

Figure 10-10 DC drive for Question 4.

5. a. Complete the wiring of the armature circuit for the nonregenerative DC drive circuit shown in Figure 10-11.

   b. In this circuit, braking is achieved by dissipating the braking energy through the DB resistor as _____.

Figure 10-11 Armature circuit for Question 5.

6. a. Complete the wiring of the armature circuit for the regenerative DC drive circuit shown in Figure 10-12.
   b. In this circuit, braking is achieved by conducting the braking energy back to the _____.

Forward Bridge   **Reverse Bridge**

Figure 10-12 Armature circuit for Question 6.

7. An overhauling load
   a. Results in generator action within a motor
   b. Occurs whenever the inertia of the load is less than that of the motor rotor
   c. Occurs whenever the inertia of the load is more than that of the motor rotor
   d. Both a and c
   Answer _____

8. Programming parameters associated with DC drives are similar to those associated with AC drives. (True/False) _____

9. Which of the following DC drive parameters sets the time it takes to change speeds from one setting to another?
   a. Acceleration time
   b. Deceleration time
   c. IR compensation
   d. Maximum speed
   Answer _____

PART 4   Programmable Logic Controllers (PLCs)

1. A programmable logic controller (PLC) is an industrial-grade _____
    capable of being programmed to perform control functions.

2. PLCs are the most widely used process control technology.
    (True/False) _____

3. Identify the components of the block diagram of a PLC shown in Figure 10-13.

    a. _____   _____   _____

    b. _____   _____

    c. _____   _____

    d. _____   _____

    e. _____   _____

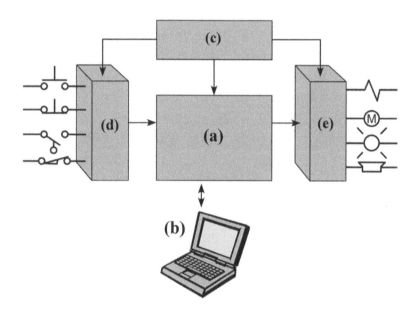

Figure 10-13 Block diagram for Question 3.

4. The programming device must be connected to the PLC at all times in order to
    run the program. (True/False) _____

5. PLCs were originally designed as replacements for

    a. Microcomputers

    b. Relay control panels

    c. Analog controllers

    d. Digital controllers

    Answer _____

6. The central processing unit or processor of a PLC

   a. Looks at the inputs, makes the decision provided by the program, and sets the status of the outputs

   b. Looks at the outputs, makes the decision provided by the program, and sets the status of the inputs

   c. Serves only to store the program in memory

   d. Understands only ladder logic

   Answer _____

7. The power supply required to operate the PLC processor and I/O modules units is

   a. Low-voltage AC

   b. High-voltage AC

   c. Low-voltage DC

   d. High-voltage DC

   Answer _____

8. PLC input modules

   a. Receive signals from field devices

   b. Convert input signals to logic signals

   c. Provide isolation between field devices and PLC circuitry

   d. All of these

   Answer _____

9. PLC output modules convert signals from the CPU that are used to control output field devices. (True/False) _____

10. Which of the following is *not* a function of the PLC programming device?

    a. To enter the user program

    b. To change the user program

    c. To run the user program

    d. To monitor the user program

    Answer _____

11. A nonmodular or fixed PLC consists of a _____, _____, and _____ in a single unit.

12. The chassis of a modular PLC is divided by compartments into which the separate _____ can be plugged.

13. A _____ is a user-developed series of instructions that direct the PLC to execute actions.

14. The relay ladder logic (RLL) programming language is based on electromagnetic relay control. (True/False) _____

15. A relay ladder logic program graphically represents rungs of _____, _____, and _____, _____ blocks.

16. Answer the following with reference to the PLC motor start/stop program shown in Figure 10-14.

   a.   What is the address associated with each of the following field devices?

      (1) Stop button _____   (2) Start button _____

      (3) Starter coil _____   (4) Starter coil contact _____

   b.   What is the instruction associated with each of the following field devices?

      (1) Stop button _____   (2) Start button _____

      (3) Starter coil _____   (4) Starter coil contact _____

   c.   What is the state (true/false) of each of the following when the motor starter coil is deenergized?

      (1) I1 _____   (2) I2 _____   (3) I3 _____

   d.   What is the state (true/false) of each of the following when the motor starter coil is energized?

      (1) I1 _____   (2) I2 _____   (3) I3 _____

Figure 10-14 Start/stop program for Question 16.

17. The addressing format used for PLCs is the same for all manufacturers. (True/False) _____

18. In RLL programs outputs are represented by _____ symbols.

19. The term _____-wired refers to logic control functions that are determined by the way the devices are interconnected.

20. From a safety standpoint, only normally _____ push buttons should be utilized for motor stop buttons.

21. Complete the wiring for the PLC shown in Figure 10-15 programmed for a motor start/stop circuit similar to that shown in the text, using the following addresses:

    Normally open start button—I2

    Normally closed stop button—I6

    Starter coil—Q4

    Normally open starter contact (M1)—I8

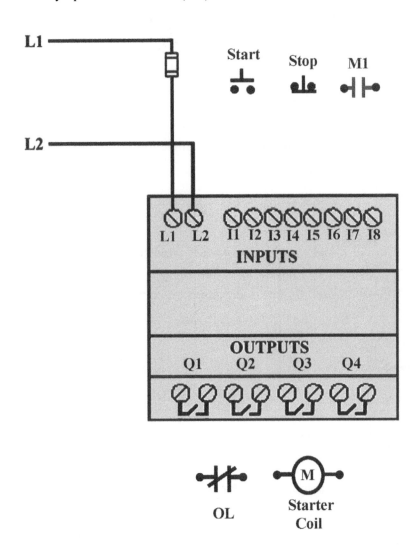

Figure 10-15 PLC diagram for Question 21.

22. One additional start and stop button is to be added to the original PLC motor circuit containing remote standby and run pilot lights outlined in the text and shown in Figure 10-16. Complete the ladder logic program for these changes.

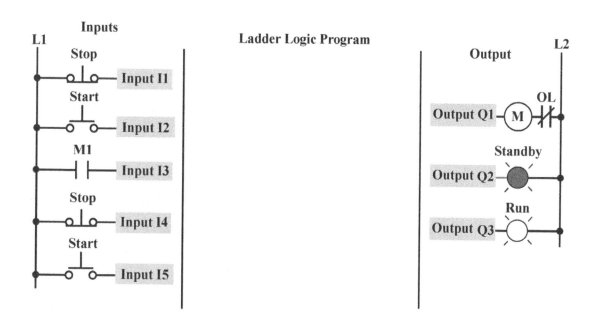

Figure 10-16 PLC motor circuit for Question 22.

23. PLC timers

a. Are input instructions

b. Are output instructions

c. Provide the same function as electromechanical timing relays

d. Both b and c

Answer _____

24. Which of the following PLC timers is used to delay the start of a machine process?

a. Retentive timer

b. On-delay timer

c. Off-delay timer

d. On/off timer

Answer _____

25. Which of the following changes state whenever the accumulated value of the PLC timer reaches the preset value?

a. Enable bit

b. Done bit

c. Time base

d. Timer number

Answer _____

26. Answer the following with reference to the diagram of the PLC timer program shown in Figure 10-17.

a. This timer would be classified as an _____-delay type.

b. The time-delay period for this timer is set for _____ seconds.

c. Pilot light output B will be on at all times that the switch is opened. (True/False) _____

d. Pilot light output C will come on when the switch is closed and the time-delay period has elapsed. (True/False) _____

e. The timer's accumulated value resets to zero any time the switch is opened. (True/False) _____

Figure 10-17 PLC timer program for Question 26.

27. The correct address for a pilot light connected to terminal 5, slot 3, of a single-rack Allen-Bradley SCL 500 controller would be

   a. O:3/5

   b. O:5/3

   c. I:3/5

   d. I:5/3

   Answer _____

28. Programmed counters can

   a. Count up

   b. Count down

   c. Be combined to count up and down

   d. All of these

   Answer _____

29. Answer the following with reference to the diagram of the PLC timer counter program shown in Figure 10-18 (p. 230).

   a. This counter would be classified as an _____ counter.

   b. The counter is set to actuate the done bit instruction whenever a count of _____ is reached.

   c. The on-to-off transition pulses of push button _____ are counted by the counter.

   d. What would be the state of the pilot lights (on or off?) after the following sequence:

   • PB2 momentarily actuated

   • PB1 actuated on to off eight times

   • PB2 momentarily actuated

   • PB1 actuated on to off two times

   (1) Red pilot light _____

   (2) Green pilot light _____

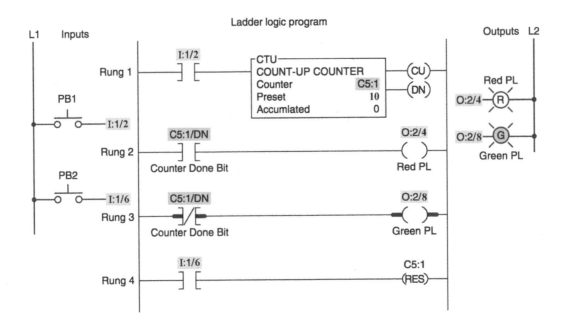

Figure 10-18 PLC timer counter program for Question 29.

30. Complete the wiring for the PLC shown in Figure 10-19 using the addressing and program of Question 29.

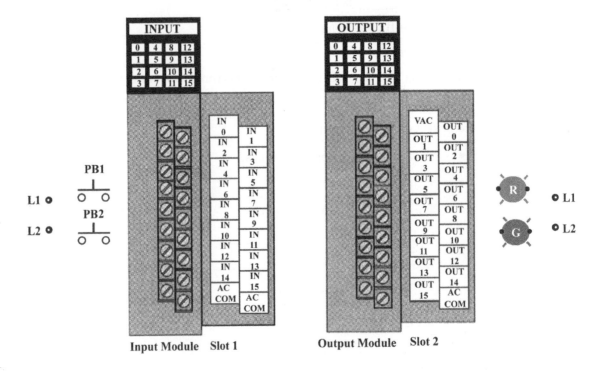

Figure 10-19 PLC diagram for Question 30.